V

The Brain Machine

The Brain Machine

The Development of Neurophysiological Thought

MARC JEANNEROD

Translated by
David Urion, M.D.

Harvard University Press
Cambridge, Massachusetts, and London, England
1985

Copyright © 1985 by the
President and Fellows of Harvard College
All rights reserved
Printed in the United States of America
10 9 8 7 6 5 4 3 2 1

Originally published in 1983 by
Librairie Arthème Fayard, Paris, as
Le Cerveau-Machine: Physiologie de la Volonté
by Marc Jeannerod

This book is printed on acid-free paper,
and its binding materials have been chosen
for strength and durability.

Library of Congress Cataloging in Publication Data

Jeannerod, Marc.
The brain machine.

Translation of: Le cerveau-machine.
Bibliography: p.
Includes index.
1. Brain—Research—History. 2. Neurophysiology—
History. I. Title.
QP376.J413   1985   612'.82   85-7597
ISBN 0-674-08047-5

Preface

The brain controls our actions and our behavior. Over the course of three centuries, along an astonishing scientific path, experiments and hypotheses have succeeded one another in an attempt to explain the nature of this control. In turn, the propagation of the nervous influx, the relationships between the brain and the organs to which it responds—and which respond to it—and the identification of the cerebral localization of the motor system have each benefited from facts gleaned through observation, animal experimentation, and analysis of clinical material. At each stage in this progression, *movement* has been considered a special index: whether a reflex sending back the image of excitation, a response to electrical stimulation of the brain, or an element of diagnosis in a neurologic examination, it is movement that has been observed, measured, and recorded.

Tracing the history of the physiology of motor production is thus interesting in and of itself, as a way of understanding how our current knowledge of the nervous system was progressively organized. However, at whatever level one approaches it, neurologic discussion is more than a specialized discussion regarding the function of an organ. Movement is the visible climax of a process which transforms energy from outside the subject. Moreover, when it is spontaneous, it reveals the internal state of the subject who produces

v

it—that subject's intent, will, and power to act. Movement thus becomes an act of creation, of communication, of transformation of the environment. From this perspective, how could one envisage a true physiology of the subjective? Up to what point can one (or should one) be objective? Ever since Descartes, no study of the cerebral organization of motor function has been able to avoid the debate surrounding these questions. The answers offered by physiologists, neurologists, and psychologists, as well as the controversies those answers produced, are also fascinating contributions to the problem of the relationships between the body and the soul, the brain and the mind, matter and energy.

By its role as mediator between the organism and its environment, movement—whether reflex, automatic, or spontaneous—also belongs to the biologic discussion in its broadest sense. On the one hand, motor function can be thought of as purely an instrument of adaptation to the environment, of submission to the natural order. This point of view obviously singles out the reflex as the functional modality that explains behavior. Inversely, the reaffirmation of the physiologic autonomy of another form of motor activity, spontaneous movement, leads to a conception with a radically different emphasis: that of an organism which structures the environment around it, imposing its own organization. Spontaneous movement thus becomes the means by which the organism interacts with its environment, and with whose aid the subject constructs its representation of the real.

I would like to thank Professor Richard Held (Cambridge, Mass.) for giving me a copy of F. Fearing's indispensable book, and Professor Charles Phillips (Oxford) for giving me a copy of E. G. T. Liddell's book. Jacques Mehler, Claude Debru, and Jean Bullier read the first versions of my work and graciously offered their advice. Finally, M. Althabe and F. Girardet kindly saw to the typing of the manuscript.

Contents

The Brain Machine

1

The Natural History of the Soul

The acquisition of any new knowledge, and even the mere observation of facts themselves, are often subject to the conceptual and historic setting. The evolution of an idea suddenly reveals the existence of a shifting zone, where knowledge had previously seemed stable and secure. This certainly characterizes the evolution of ideas regarding the structure and function of the brain which occurred during the end of the seventeenth and the first half of the eighteenth centuries. From this era we can date the concepts which serve as the foundations of modern anatomy and physiology. The teachings of the ancients on the physical seat of the soul, even its cerebral localizations, which had lasted until then, were abandoned. Galen's descriptions of the brain, certainly incomplete but by and large correct, were no longer afforded unlimited confidence. Everything had to be rediscovered or reexplained. Just as E. B. de Condillac had described man as devoid of knowledge, constructed little by little by his environment, the scientist of the eighteenth century was a *tabula rasa* who allowed himself to be guided by observation.

Unless it is carefully prepared in the laboratory, the brain defies anatomic description. In its fresh state, it collapses, deforms, and

deteriorates rapidly. Only the most obvious morphologic characteristics can be observed consistently, and thus it is not surprising
that the ventricles, those well-demarcated cavities within each cerebral hemisphere and in the upper brainstem, first struck early observers. From them, the notions of receptacle, reservoir, and
circulation arose. The content of the ventricles (air, according to
Galen; fluid, according to the authors of the Renaissance) thus
appeared to be a plausible vehicle for transporting the "animal
spirits," a term which persisted until the middle of the eighteenth
century as a designation for the unit of nervous transmission. As
for the cerebral substance itself, the cortex or grey matter was
thought to secrete the fluid, like a gland, while the white matter,
with its multiple canals, served to transport the fluid to nerves,
which in turn transported it to muscles.

Descartes's representation of cerebral function, without a doubt
the most complete at the end of the seventeenth century, took this
hydraulic model into account.[1] According to Descartes, sensory
stimulation produced a flux of the animal spirits contained in the
heart and arteries. The heart then pushed the spirits into the cerebral cavities, much as the pumps of an organ push air into its pipes.
Pores permitting the passage of the animal spirits were opened in
a selective manner by the stimulus which produced the flux, much
as the fingers of the organist distribute the air from the bellows
into several pipes at once. The animal spirits inflated the cavities,
and pulled on the little tubes which were attached to them, just as
the wind can "fill up the sails of a ship, and put all the lines attached
to them under tension."[2] It is this action of the animal spirits that
made the little tubes permeable; by contrast, after death the brain
collapsed and fluid could no longer circulate. In the conception of
Descartes, any action which occurred in response to a stimulus was
due to the intervention of the pineal gland. The seat of imagination
and common sense, the pineal gland received information about the
external object that served as the source of the stimulus. An image
of the object was formed on the pineal gland, which itself then
leaned toward the side bearing the image. In so leaning, the pineal
gland pushed on certain tubes and closed them, while it distended
others which in turn became permeable. The animal spirits then
flowed out to the appropriate muscles and caused them to contract.

The analogies of the pump of a pipe organ or the inflated sail, the confluence of the circulatory and nervous systems, the cavities, the open and closed tubes, all converged to support the notion that neural information was transmitted via a circulating fluid in the nerves. Even though the Cartesian conception of cerebral function anticipated certain aspects of modern physiology, it appears *a posteriori* to have rather weak explicative value. In the domain which concerns us, the production of movement, this model is limited to explaining the unleashing of an action by an external stimulus. But

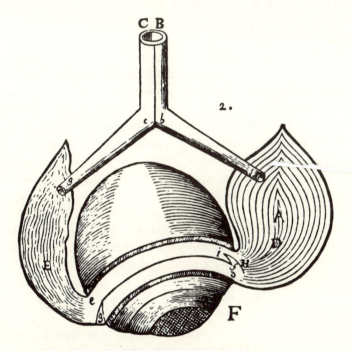

Figure 1 Representation of the command for muscular contraction according to Descartes (1664). Animal spirits arrive by nervous tubes *B* and *C*. In the case presented here, animal spirits inflate muscle *A*, while muscle *E* relaxes. A movement (here, of the ocular globe) is produced, toward muscle *A*. In order to explain this reciprocal innervation mechanism by the hydraulic model, Descartes had to postulate the existence of a complicated system of valves which allowed for the evacuation or retention of animal spirits. In the case presented here, valve *H* closes, which allows muscle *A* to be filled, while muscle *E* can freely empty via evacuation canal *e*.

how could one conceive of the autonomous functioning of such a system of pumps and valves? How does one take into account the appearance of a spontaneous movement, one unrelated to an external event? At that time, the problem of voluntary movement was not yet posed in those terms; it was considered part of those phenomena which came from the self, and for which no explanation was really sought.

If the hydraulic model of the nervous system captures our attention, it is less for its interest in its own right than for the manner in which it was refuted. Experiments showing its incompatability with observation, though very simple, did not convince the scientific community of the era. The theory that replaced it, which proposed an electrical propagation of the *vis nervosa* (nervous force), found itself up against numerous difficulties.[3] By the end of the seventeenth century it was appreciated that nerves were not, in fact, tubes. Around 1674 A. van Leeuwenhoek, who had a microscope at his disposal, examined the nerves of a calf but was unable to conclude in any decisive manner for or against the tubular nature of nerves. However, A. Monro noted that no liquid ran out when a nerve was sectioned, nor did a ligated nerve swell behind a constriction. Thus, he felt that nerves were simple cords, and not tubes containing a circulating liquid. Moreover, it was noted that a muscle contraction occurred without producing a change in volume. J. Swammerdam demonstrated this by placing a sectioned frog's leg, isolated with its nerve (a nerve-limb preparation), in a vessel filled with water. Pinching the nerve, he brought about a contraction of the leg; this did not produce any change in the level of the water. Monro had indeed thought of attributing the transmission of the *vis nervosa* to an electrical phenomenon, but wondered how electricity could selectively propagate itself along a nerve without diffusing into the surrounding tissue, since nerves were devoid of any insulating substance. At the time, the criteria required to confirm the electrical nature of a phenomenon were quite strict: it was considered necessary to produce a spark. Around 1730 P. van Musschenbroek, the inventor of the Leyden jar, thus concluded that electricity could not be involved in nervous transmission, and stated that the nervous fluid was propelled through the nerves by a systole of the brain, just as the blood was propelled through the arteries by the heartbeat.

Support for the electrical model would require the long development of a new concept that owed nothing to the hydraulic model. Noting that isolated muscle contracted under the influence of simple mechanical excitation, F. Glisson in 1677 used the term "irritability" to designate this property. The nerve itself was also "irritable," since pinching it provoked the contraction of an attached muscle. A. von Haller, who from 1752 onward became the veritable propagandizer of this idea of "irritability," distinguished between "irritable nerves" (those in which mechanical or chemical excitation produced a muscular contraction) and "sensitive nerves" (in which excitation produced no contraction). It was the latter, he thought, which transmitted impressions to the soul. It should be noted that von Haller invoked electrical transmission to explain the phenomenon of irritability but, similar to Monro, he stumbled on the question of insulation of conduction, and, for lack of a better explanation, finally resorted to the theory of nervous fluid as the vehicle for *vis nervosa*. Animal electricity, only suspected through the notion of irritability, was not discovered until 1791, by L. Galvani in Bologna.

Galvani first showed that the nerve-limb preparation of the frog (the same preparation that Swammerdam had used) was sensitive to electricity: the limb contracted if the nerve was put in contact with two different metals (iron and copper). Moreover, Galvani discovered that the limb also contracted when the nerve was put in contact with another freshly cut muscle. Irritable tissue therefore not only was sensitive to electricity but also produced it. These spectacular results were the object of a long controversy with A. Volta, which continued long after the death of Galvani. Not until 1850 were the Galvanic phenomena not only reproduced (this had already been done by Baron von Humbolt in 1797 and by C. Matteucci in 1842) but finally integrated into an explanation of nervous function. In 1848 E. du Bois-Reymond assimilated the nervous principle, the *vis nervosa*, into the principle of electrical propagation. Nervous influx had definitely supplanted animal spirits.[4]

Parallel to this fundamental work on isolated nerve and muscle (which would lead to the elucidation of the mechanisms of synaptic transmission by the English school at the beginning of the twentieth century), another area of research was developing. More global in its inspiration, at the same time more neurologic and more anatomic, this research would lead to the discovery of cerebral localization and the role of the cortex in the mechanisms of language, perception, and action. This distinction between two complementary approaches to the study of function in the nervous system, though somewhat schematic, nevertheless continues today: on the one hand, the study of cellular and subcellular mechanisms; and on the other, the study of relations between the structure of the brain and behavior. We will utilize the second approach, for the most part, because it seems to be the only one able to answer the questions posed by the production of movements and the organization of action.

As we have seen, the understanding of the anatomy of the brain was still poor during the seventeenth century. The techniques of the time were rudimentary—no microtome, no staining, no microscope existed. Fixation, other than oil immersion, was not done, which made dissection laborious and uncertain (alcohol fixation was not introduced until 1785, by F. Vicq d'Azyr). In addition, the hydraulic theory had given undue prominence to certain morphologic aspects of the organ, in particular its cavities. The work of T. Willis on the anatomy of the brain (published in Latin in 1664 and translated into English in 1683) constituted a new point of departure. The ventricles lost their dominant position: Willis thought they were, in all probability, merely organs of purification and excretion. The most revolutionary aspect of this work does not lie in the renewal of the description of the brain per se, however, but rather in the recognition of anatomic differences to which one might attribute functional differences. The cerebral cortex was thus attributed a role in memory; the corpus callosum, which joined the two hemispheres, was given a role in imagination. The cerebellum became the seat of involuntary movements, and the striatum the seat of the *sensorium commune*, where all the impressions coming from the various sensory organs converged. Finally, the quadrigeminal bodies were implicated in sensations arising from the heart and the viscera. Even if Willis's model of nervous function was still strongly impregnated with hydraulic theory, his work showed the importance of a preliminary, objective anatomic description to any study of the relationship between the cerebral structures and given functions, such as the production of movement. Anatomy is the key to physiology, though it is a dead science without physiology.

In 1683 physiology was only in its infancy. In fact, the debate which began with Descartes and Willis on the nature of the connection between the soul and the brain is not really a physiologic one. The soul of reason, said Descartes, "cannot in any way originate from the power of matter." It is lodged in the body (in the pineal gland), but not merely like "a pilot in a boat." So that man might have feelings and appetites, the soul had to be "joined and united" with its body. But whatever the intimacy of the connections between them, soul and body remained separated by a difference of nature, of substance. No direct route leads from thinking to being, save by God; our understanding of external objects, and our actions upon them, are not possible except by this route. The soul, then, must be "expressly created," and reason is in us from the beginning, innate.[5] The animal has no soul, according to Descartes; it is a machine, an automaton: "If there were machines that had the organs and the features of a monkey, or some other animal without reason, we would have no means of recognizing that they were not of the same nature as these animals. However, if there were machines that resembled our bodies, and which imitated as many of our actions as might morally be possible, we would always have two certain means for recognizing that they were not real men": they would not have language or consciousness.[6]

The dualist line of thinking was thus solidly established, and would become official doctrine for a long time. Those who strayed found themselves brought to inquest, often banished, their books seized and burned. The general tendency was, therefore, to prudence; the researcher confined himself to observation and classification, and kept his interpretations to himself. Did not attributing distinct intellectual functions to different parts of the brain challenge the notion of the one and indivisible soul? As we will see at length below, the segmentation of the psyche into dissociable functions was a necessary condition for breaking this concept of the soul which blocked any objective study of neural mechanisms. Moreover, any later attempt at reunification of the elementary psychic functions (or certain of them) as the core of some higher level or more global psychic entity would inevitably appear to be the recrudescence of dualist or spiritualist thought.

Before disappearing entirely, however, the notion of the soul underwent a subtle evolution. Stepping down from its pineal throne, the soul was introduced into the organs and movements. No longer arbitrating the whole, it was decentralized into each of the parts of the whole. While remaining immaterial, and without a relationship in substance with the body which it animated, the soul became the "vital principle," the energy present in each action or each manifestation of the living being; without this diffuse spiritual energy, the different parts of the body would be passive. This point of view, supported in a rather dogmatic fashion after 1708 by G. E. Stahl, seemed a regression when compared with the theories of Descartes. The Cartesian dualism constituted an explanation of "last resort," where the soul was certainly necessary for the functioning and cohesion of the whole, but left the field open for the existence of physiologic mechanisms. For Descartes, the mechanisms of our movements were of the same "machine-like" nature as those of an animal-machine. By contrast, the soul according to Stahl had become a general explicative principle which would account for the details of any movement, just as in chemistry phlogiston could serve as a sufficient explanation for any reaction. This was no longer dualism; this was animism.

This theory had a wide following, above all because it represented a rather coherent alternative to the growing influence of the "iatromechanicists," whose theories, while not quite materialist, were nevertheless rather mechanistic.[7] Could one not in fact demonstrate that the soul is not localized solely in the brain, but that it is everywhere? The physiologic experimenters rediscovered that a decapitated frog continued to move when stimulated. The brain was therefore not necessary for the movement of the extremities, which contradicted the notion of free will and volition localized in the head. As R. Whytt stated in 1751, the soul continued to act on the separated parts even after decapitation. This evolution of ideas is particularly apparent when one reads J. O. de LaMettrie's pam-

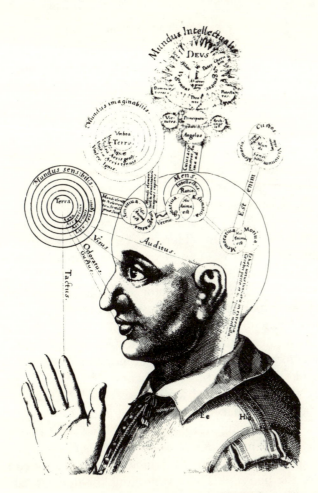

Figure 2 Diagram constructed in 1619 by the alchemist N. Fludd to explain mental function. The qualities of objects in the *sensory world* were detected by the sensory organs. In the interior of the skull, transformations occurred which resulted in knowledge, memory, etc. This representation of cerebral mechanisms in the form of three "cells" which communicate among each other had become classical by the end of the Middle Ages, and would remain so until Descartes.

In the "exterior," in an indeterminate space which the large script indicates, is the *intellectual world* dominated by God, angels, and virtues, and the *imaginary world*, where the earth's shadow hid itself. The entities acted directly on the elements of the psyche, and by combining themselves with elements of the sensory world carried by the sensory organs, constituted the soul. The entirety of the process resulted in making *motives*, which communicated with the rest of the body by the intermediary of the spinal cord.

phlet *L'Homme-machine* (*Man-Machine*). In this text, more convincingly than in his weighty *Natural History of the Soul* (1745), LaMettrie enumerates the arguments which demonstrate definitively (to his eyes) that conscience, remorse, and thought are made of the same substance as the organs. Isolated from the body of an animal, the muscles contract when stimulated and the bowels continue to manifest peristaltic movements for long periods of time. The heart, removed from an organism, will continue to beat, as experiments with frogs, chickens, rabbits, and even (as F. Bacon claimed to have observed) man showed. Moreover, pieces of heart, and not just entire hearts, continue to move. "Here are far more facts than needed," he concluded, "to prove in an incontestable manner that each small fiber or part of any organized body moves itself by a principle unique to itself."[8] This was a "motor principle," common to animals and man, an innate force which resided in the "parenchyma," the noble part of the organs.

However, LaMettrie went further, since he felt that this independent force contained in the parenchyma could only produce machine-like, or automatic, movements. This force in the organs was only due to "little subservient springs." "There is another, more subtle, more marvelous one that animates them all,"[9] a principle that was "inciting and impetuous," the source of our feelings and thoughts. "This principle exists, and has as its seat the brain, at the origin of the nerves, by which it exercises its empire over the rest of the body."[10] There is nothing "behind" or "above" this principle: "By this all that can be explained is explained." LaMettrie's thoughts, expressed so unequivocally, were scandalous. His *Natural History*, published in Paris, was condemned to be burned, and the author himself had to go into exile in Leyden, Holland. There, however, the clergy and orthodox thinkers rose up against him after the publication of *Man-Machine*, forcing him to flee once again, this time to the court of Frederick II, where he found refuge.

The novel (and, for its time, subversive) character of LaMettrie's work resided in its attempt to unify man in the interior of his body. It used concepts taken from Descartes and Stahl, considered perfectly orthodox, but extended and transformed them. The innate and unique "soul" of Descartes became the "principle of action" of the brain, "because the brain has its muscles for thinking, just as the legs have theirs for walking."[11] This was the first openly materialist formulation of the problem of the soul, a way of thinking which would, in a few years, become generally accepted and commonplace. As for the ubiquitous soul of Stahl, LaMettrie took up this idea as well: his "motor principle" (the innate force which belonged to the parenchyma) and the more or less lively springs which the parenchyma contained, were quite similar to Stahl's soul. In my opinion, however, these concepts are even closer to the idea of "irritability," which was being developed at this time. In fact, LaMettrie dedicated *Man-Machine* to von Haller, who refused to accept the dedication, not out of disagreement with LaMettrie's thesis but out of spite, believing himself plagiarized.

Around 1750 two new currents of thought, at first underground, were now converging in the open, soon to provoke an ideologic upheaval. While materialism was settling its score with religion (a bit noisily, and not without damage), empiricism was combatting the notion of innate ideas which, as we have seen, was one of the central aspects of the conception of the soul in Descartes's time. Born in England with John Locke, empiricism put aside the metaphysics of the soul in order to concentrate on its "historic" aspects (today, we would say its ontogenetic aspects). The way in which knowledge organizes itself illuminates the relationship between the soul and the body. Aristotle's axiom, adopted by all the "sensualists," that "nothing reaches understanding which did not already exist as sensation," underlines the idea that sensory input is the foundation of all knowledge. At the extreme, as Locke's successors thought, there should be no difference between the external experience (sensation) and the internal experience (reflection), because to accord any autonomy to reflection, when compared with sensation, would end by restoring dualism. More than there not being innate ideas, there could be no innate operations of the soul which could permit the analysis and judging of these ideas. Condillac, above all others, should be credited with having extended the critique of Locke in the most radical fashion, and to have denied the existence of a reserved sector where reflection could be found "already there" in some way, before the arrival of the first sensations. For Condillac, all the attributes of the soul took their origin from sensation itself, and through it, from the world around us.

As D. Diderot said, "That which we call *linking of ideas* in our understanding is only the memory of the coexistence of phenomena in nature; that which we call *consequence* in our understanding is nothing but the memory of the sequence or succession of effects in nature."[12]

Although this debate, far from being closed, appeared in each era under a new guise, it took a polemical form in the second half of the eighteenth century. Empiricism had become a political argument, a reinforcement for the materialists, this time from the camp of the philosophers and no longer from the metaphysicists. Since the soul was not created but built itself progressively from experience and education, an extension of the same logic would give the soul a material site, a mechanism, a functioning. With this reasoning one could extract oneself from the philosophic and scientific impasse which held the soul to be a substance different from the body, and therefore a principle outside the order of things, outside of causality. Thus LaMettrie had initiated a school of thought. In 1770 his theses were taken up in *The System of Nature* by Baron d'Holbach, a militant materialist who, nevertheless, published under a pseudonym. As a sign of the times, the book was reprinted six times in ten years. The identity of the soul and the brain was clearly proclaimed. "Those who distinguished the soul from the body, have only seemed to distinguish the brain from itself. In fact, the brain is the common center, where all the disparate nerves from all the parts of the human body end and join one another. It is with the aid of this internal organ that all the operations attributed to the soul are accomplished. It is impressions, changes, and movements communicated through the nerves which modify the brain and, in consequence, it reacts, or acts upon itself, and becomes capable of producing, in itself and of itself, a great variety of movements which are designated by the name *intellectual faculties.*"[13]

The participation of the brain in the genesis of intellectual or mental phenomena was scarcely questioned after this time. Whatever the theory envisaged to explain the relationship between the soul and the body—between the physical and the moral, as P. J. G. Cabanis expressed it—no one would again seriously doubt the participation of the brain in such a process. According to J. C. Reil, who combatted Stahl's ideas at Halle (in the very city where the master taught his vitalist thesis), the brain is the "mysterious nuptial bed where the body and the soul celebrate their orgies." As inheritors of the Enlightenment, the ideologues of the post-revolutionary period (with Cabanis at their head) claimed that the soul is not independent of the body, and that consciousness, the topmost level of mental activity, depends upon the functioning of the brain. Thought is thus nothing but a secretion, or the result of a digestion. Cabanis said, "In order to give an accurate idea of the operations which produce thought, one must consider the brain like a particular organ, specially designed to produce thought, just as the stomach and the intestines contribute to the operation of digestion . . . Impressions, arriving at the brain, make it spring into action, just as food, falling into the stomach, excites the latter into the most abundant secretion of gastric juice . . . The sole function of the one is to perceive each particular impression . . . just as the function of the other is to act upon nutritive substances."[14]

Figure 3 Dissection of a brain by H. Mayo (1827). The fiber bundles coming from the spinal cord, and traveling to the cerebellum and cerebral hemispheres, can be clearly seen. Mayo also distinguished the bundles connecting the different convolutions of the same hemisphere.

Further, "We conclude, with the same certainty, that the brain digests, in some fashion, impression, and that it organically produces the secretion of thought."[15] Cabanis had, moreover, performed physiological experiments on animals by locally stimulating the brain. He noted that he could produce convulsive movements of different muscle groups by stimulating different zones of the brain. This observation, of great importance in understanding the mechanisms of movement, would not be studied in depth until much later, around 1870.

Understanding of the anatomy and functioning of the brain were making rapid progress. Anatomy was dominated by the personality of F. J. Gall, who had been exiled from Vienna, where his ideas had not been well received; he then passed through Halle, where Reil was established, and finally settled in Paris around 1807. Gall had perfected the classical technique of R. Vieussens, which consisted of dissecting the brain by following fiber bundles up to the cortex. He was thus able to show that the cerebral convolutions, which up until then had been considered vague "enteric processes" resembling the loops of the intestines, in fact constituted a folded but continuous mantle, where fibers had their endings. He also managed to identify certain convolutions, numbering them and indicating "the locations where instincts, sentiments, desires, talents, and, in general, the moral and intellectual forces were exercised." Gall indicated the "organs" which, according to his phrenologic system, corresponded to the numbered convolutions in his atlas of the cortex. Thus was born cerebral localization, a central concept of modern neurology and neuropsychology. None of the specific localizations which Gall proposed have been retained, however; his list of the intellectual functions was purely arbitrary. Much later, others (particularly J. B. Bouillaud and P. Broca), applying knowledge of the cortical anatomy and observation of the clinical effects of localized lesions in man, were able to establish the correlation between lesion and symptom, and then attempt to establish the correlation between localization and function. Even if Gall's scientific anatomy was later to degenerate into phrenology and cranioscopy, these failings in no way detract from the definitive nature of his concept of cerebral localization.[16]

The alternative to the materialism of the ideologues was represented by the vitalist thesis. The vitalism of the nineteenth century was a difficult conglomeration to sort out, standing for more than it really was in the eyes of those who opposed it. At first a substitute for the soul, the "vital force" progressively became tied to notions of organism and organization of life. Life became a principle of cohesion, "the collection of functions which oppose death" (X. Bichat) and which allow an escape from the determinism of the physical. "An immense gulf," Bichat said, "separates physics and chemistry from the science of organized bodies, because an enormous difference exists between their laws and those of life. To say that physiology is the physics of animals is to give an extremely inexact impression. I would do as well to say that astronomy is the physiology of stars."[17] Vitalism, as it was represented in France at this time by the theories of the Montpellier school and Bichat, is thus the simple recognition of the originality of the existence of life with respect to the existence of matter. Life is something more. Like many provisional theories in the history of science, vitalism represented a need for an explanation that the subsequent development of the discipline would render obsolete. In physics, it was a common practice to settle a controversy about a new or unexplained property of matter by giving it a name. Later, in 1878, Claude Bernard would find the *raison d'etre* of the vital force in the role of subordinating elements of an organism to the whole; life thus became a self-regulated mechanism, maintaining itself by the constancy of the "milieu intérieur" of each living organism. In the early 1800s, however, vitalism was assimilated into the spiritualist/dualist current. Even though Bichat's theories were unequivocal, their binary nature could superficially create the illusion of dualism. Bichat distinguished between organic life and animal life, each composed of two orders of functions. Through "organic life," the animal "ceaselessly transforms molecules of related compounds into its own substance, and then rejects those molecules when they become heterogeneous to it."[18] It was thus a "current of matter" which "composed" and "decomposed" the organism. "Animal life," on the other hand, could be considered as a flux of information. This current first established itself in a "passive" fashion from the exterior of the body to the brain, then from the brain to the effector organs: the animal thus became "active" in the "successive actions of the brain, from which volition is born as a result of sensation."[19] This active flux, this motility, is ruled in turn by two orders of phenomena which come, respectively, from animal and organic life. "Animal contractility," "essentially submitted to the influence of the will, has its principle in the brain, and receives from it the irradiations which put it in play, and ceases to exist as soon as the organs where one observes it no longer communicate with it via the nerves"; "organic contractility," on the other hand, "independent of a common center, finds its principle in the moving organ in question, and is not subject to voluntary acts."[20]

Perhaps in Bichat's "organic life" one can find a remnant of Stahlian vitalism, against which the materialists arose to wipe out any trace of the irrational in the description of the function of the organism. Until around 1850, one could still find descendants of Stahl in Germany. J. Müller, the influential physiologist, opposed certain of his students and questioned their results when they did not conform to the vitalist principles he espoused. For example, H. von Helmholtz had developed an ingenious technique which permitted him to measure the velocity of propagation of nervous influx in the frog. Measuring the time which elapsed between electrical stimulation of the nerve and contraction of the muscle, he estimated that this velocity was about thirty meters per second. For Müller, however, von Helmholtz should have "normally" found no delay, since in vitalist terms the command for contraction was not supposed to be distinct from the contraction itself. By the end of his life, Müller had repudiated the vitalist theory and even burned some of his older work as a sign of renunciation.

That term "vitalism" included several theoretical positions, of which certain ones ended in a scientific deadend, is not in doubt. It was, however, over this uncertain terrain that the first physiologists set out. From 1809 on, F. Magendie (who admitted to having believed, for some time, in vitalism, a religion which gave him a "consoling belief") retracted his previous statements and affirmed his membership in the materialist camp: "The brain is the organ of thought. Any number of observations and experiments prove this." These experiments were the ablation of localized parts of the brain in different animals, and had been carried out by L. Rolando in Turin and J. J. C. Legallois, Magendie, and P. Flourens in Paris.

Physiology, in its infancy, was thus a dynamic and functional counterpart to anatomy: by making a given function disappear, a lesion
allowed one to determine by default the role of the lesioned structure. The way was open for the anatomic-clinical method in neurology. From 1822 onward, Flourens reported the effects of a long
series of ablations, usually carried out on birds. The ablation of
the cerebral lobes (the most anterior part of the brain which, in
mammals, contains the cerebral cortex) produced a characteristic
effect: loss of spontaneity. The animal sat immobile, but remained
capable of movement when the examiner stimulated it. The cerebral
lobes, Flourens concluded, were therefore not the site of the immediate cause of muscular movements nor the site which coordinated the movements of walking, flight, etc. What seemed to be
lost after this type of lesion was *spontaneous* movement, those movements which expressed the "self will" of the animal. The cerebral
lobes were thus the exclusive site of "volition." To localize an intellectual process in a defined part of the brain seemed in accord
with the materialist thinking of numerous physiologists, such as
Magendie. But when Flourens defined what he meant by "volition,"
mainly in his work of 1842, he adopted a remarkably vague formulation which seemed, in fact, to draw him away from the mechanistic view. For Flourens, the brain was simultaneously the organ
of "sensation" and of "perception," of "motion" and "will." However, he made a qualitative distinction between sensation and motion on the one hand, and perception and will on the other. Sensation
and motion were simple functions which controlled, respectively,
the senses and muscular contraction. Perception and will, which
alone were localized in the cerebral lobes, were, by contrast, indivisible and united functions of a single entity: intelligence.[21]

In allowing his thinking to evolve toward the general rather
than the particular, Flourens ended up negating his efforts to dissociate and establish a hierarchy of the functions of the brain. The
distance between the specific function of motion and the general
function of will became so great that the will was esconced in a
region inaccessible to experimentation and reason. It is this which
Bouillaud reproached in a talk given to the Academy of Medicine
in December 1825. How could one say that, in effect, the brain
exerted no immediate or direct influence on muscular phenomena?

Bouillaud no doubt would have preferred, from a physiologist like Flourens, a clearer affirmation, a position more firmly in favor of the direct control of voluntary movement by the brain. He said so himself: the brain is the "movement center," not only for the extremities but also for the eyes and the tongue, movements of which play an important role in the "mechanism of the intellectual functions." The intelligence and the will could thus direct the execution of movements. "It is only necessary," said Bouillaud, "to glance at the anatomic connections between the brain and the spinal cord to convince oneself that it [the brain] must play an important role in the diverse acts over which muscular contraction presides."[22] This nervous link could transfer to the level of the spinal cord the particular movements which each of the parts of the brain had under its control. It is thus, he continued, that one could explain partial deficits of muscular function, by "the effect of a local affection of the brain." In Bouillaud's thinking, there was no reluctance to localize functions as "intellectual" as speech or voluntary movement to certain parts of the brain (that is, the cerebral cortex). The ideas of Gall had paved the way, and Bouillaud as well as Flourens recognized their debt to him. After the great exorcism at the turn of the eighteenth century had eased everyone's mind by dissipating the spiritualist aura that surrounded Gall's ideas, they allowed a simplification of the problem of the relationship between the brain and the nervous and psychic "functions."

2

Movement as Argument and Mechanism

M ovement is the most immediate and most evident expression of life; it distinguishes the animate from the inanimate. Through movement, the physiologist penetrates the interior of a living being, toward the hidden source of action. Action reveals the "internal state" which preceded it, the "intention" of the moving organism, the goal toward which action itself is directed. This objective reality of movement, which gives it an almost tangible character as an observable and recordable event, was the obligate pathway for all physiology of the nervous system in the nineteenth century. It was not until much later that other phenomena would join movement as criteria for neural activity; they would never supplant it.

The scientific status of movement is ambiguous: as an observed fact it can be described; as a revelation of the process which produces it, it becomes an object of interrogation and experimentation beyond itself. This ambiguity renders any attempt at classification of movements difficult, since the act of classification itself implies a hypothesis regarding the nature of the phenomena so classified. The categories of movement adopted in the seventeenth century barely took the descriptive, "exterior" aspects of movement into account.

To the contrary, they reflected a theoretical attitude about the global function to which a given moment was felt to belong, about the functional and anatomic hierarchy from which it arose. In the extreme, a classification of movements subtended a classification of beings which produced them. Descartes thought that human movement is what defined man, because it revealed his soul and the superior principle which inspired it. Animal movement is similar to human movement, said G. L. Buffon, but this resemblance is external and does not allow us "to pronounce the nature of man similar to that of an animal." For that, it would be necessary "to know the internal qualities of the animal as well as we know our own."[1] The distance between animal movement and man's rational soul was thus as great as the distance between the passive vegetable and the animate animal.

The criteria first used to classify movements were based on suppositions about the greater or lesser dependence of each type of movement upon the intervention of the soul or the will. The will, in this context, was an ill-defined attribute of the soul, translating the conscious capacity of wishing into an effective execution of an action or, on the other hand, translating the capacity of wishing a change into the suspension or turning away from an initial goal. The elucidation of this term "will," and the multiple concepts it covered, without a doubt constitutes the central problem in comprehending the physiologic bases of the different aspects of movement. Is the will an integrating part of the motor function, a sort of "initial stage" for certain types of movements? Or is it only an abstraction, a state of the subject, a free will which presides over certain actions?

These questions were not yet posed during the era when the classification of movements was first emerging, a system based as much upon the teaching of the ancients as upon the work of the first physiologists. During the second half of the seventeenth century, Vieussens described "involuntary" movements, which were a "mechanical" result of the structure of the moving parts. The archetype of these movements was the heartbeat: no order from the will preceded it; it was executed without the soul's knowledge. "Intrinsic" movements were also not influenced by the will, and even acted against it. They were precipitated by an external influence, though the soul perceived them and felt them. "Mixed movements" were, in some ways, ancient voluntary movements which had become involuntary by the force of repetition. They were guided by habit and not by the will, but the latter could inhibit them or modify their execution. As for "voluntary" movements, they were defined as all movements which were none of the above, thus being delimited more by comparison and exclusion than by their own characteristics. In 1677 Glisson proposed that involuntary movement was that motor function which persisted in the decapitated animal, a motor function consisting of movements dependent upon an inferior form of perception that did not reach the spirit. This, therefore, implicitly attributed an encephalic seat to the other motor function, voluntary movement. Vieussens's anatomic hypothesis regarding the cerebral partition of movements, a more structured notion than Glisson's, testifies to the emergence of the anatomy of the brain. For Vieussens, the part of the brain situated beneath the striatum, containing the centrum semiovale, the olives, the pyramids, and the cerebellum, was the origin of involuntary movements. Other movements arose in the brain above the striatum.[2]

Thus, toward the end of the seventeenth century, a binary scheme of motor function was elaborated. Involuntary movement was located in the inferior part of the brain and descended into the spinal cord; voluntary movement was located in the superior part of the brain and in the cortex. The functional hierarchy imposed an anatomic hierarchy, by placing the noblest, most human (or least animal) motor function at the highest level, the farthest distance possible from the entrance or exit of the nervous system.

This was almost an isolation: exiled to the cortex, voluntary movement would remain an object of veneration and deference for a long time, and would almost completely escape any attempts at explanation. It was true that its existence did not have to be demonstrated: to produce a voluntary movement, was it not sufficient to *wish* it? We are in command of our muscles, we control our own actions in a manner that does not require any explanation other than metaphor. By contrast, what a discovery it was to find movements which avoid this voluntary control, and are executed unconsciously, even despite it! The classifications and hierarchies of movements therefore appear not so much attempts to rationalize and explain the "highest" categories, but rather as attempts to integrate, if possible in a familiar context, the surprising phenomenon of involuntary movement. In a way, involuntary movement was a revolutionary idea, a breach in the edifice of rational man, through which the machine-like nature of animal movement infiltrated into the terrain of the soul itself. No one thought so at the time, but it was thanks to this providential breach that the edifice would be destroyed and rebuilt.

In Willis's classification, voluntary and involuntary movements were separated and their difference explained in anatomic terms. Involuntary movements represented a category apart, consisting of such things as the pulsations of the heart, respiration, and intestinal peristalsis. The cerebellum alone was responsible for these activities, distilling its animal spirits in its cortical part and then distributing them to the organs via its medullary substance. This function of the cerebellum was common to mammals: did not the cerebella of all species resemble one another? It was otherwise in the forebrain, the seat of voluntary movement. The brain, according to Willis, consisted of the medulla oblongata (which corresponds, in today's terminology, to the brainstem and spinal cord), the cerebral hemispheres, and—at the junction of the latter with the medulla oblongata—the striatum. It was there, in the *sensorium commune*, that all sensations converged, and from there all impulses which gave birth to movements arose. According to Willis, the intensity of the impression arising from the striatum determined, in the end, what type of movement would be produced. If the impressions were weak, they did not go beyond the level of the striatum, but were reflected there and went out in the form of a movement which remained unconscious, and of which the rational soul was in no way aware. Only when the impression was stronger did it pass the striatum and penetrate the corpus callosum, where the soul could contemplate it at leisure. At this stage, reflection in the form of movement was always possible. On the other hand, if it went higher the image disappeared into the cortex, its path lost in the folds of the convolutions, becoming the stuff of memory and imagination. It could one day be revealed by another impression, an awakening which could give rise to a movement or to an affective state, a passion.[3]

Willis's description remarkably intuited the notion of a reflex, a term and concept which would be retained by later researchers. At the same time, voluntary movement, as a particular entity, disappeared in Willis's nomenclature, being reduced to a mere point on a scale of intensity of response to sensation. In other words, there still was a level in the brain (consisting of the *sensorium commune*, corpus callosum, and cortex) which could consciously refract an impression and give rise to a movement. But this movement, like all others, was a response to sensation, and served to return the energy received by the brain to the outside world. The problem of the classification of movements was thus greatly simplified, since an elementary process, based on a simple physical analogy of action and reaction, came to be the only explanation for the entirety of the motor phenomena.

We will see how this explanation was later generalized to behavior and mental activity. Reflex theory was, after all, the first reductionist conception of the relationship between the brain's activity and behavior, where a complex phenomenon (voluntary movement) was conceived of as only quantitatively (and not qualitatively) different from a more simple phenomenon (reflex movement). It is a small step from this attitude to proclaiming that the approach to complex phenomena must, of necessity, pass through simple phenomena, which are considered as complex phenomena reduced to their simplest expression. Reductionism thus can take on the added dimension of an epistemologic attitude toward cerebral function and its relationship to mental activity, which should not be confused with the technical necessity of isolating simple phenomena in order to be able to study them.

The episode during the eighteenth century of the automatons is revealing in this regard. Bringing to life, in a way, the dream of Descartes, J. de Vaucanson created several clever machines, one a flutist who was able to play several tunes, another a duck which batted its wings. The work was serious: this was no attempt to fool anyone, but rather an attempt to "obtain experimental intelligence about a biologic mechanism." All movements from the simplest to

23

the most complex, could be produced by combinations of rotating drums, pulleys, and counterweights. In this sense, the criterion of complexity would be inoperative for distinguishing voluntary movement from other types of movements, since the automatons' movements (which could be very complex) were always the product of workings common to all levels of complexity. This is precisely what Willis thought: a single apparatus or mechanism could explain everything, and movement would always remain dependent upon an external force—sensation, the origin of the process of "reflection." It is interesting to see the probable influence on Willis's thesis of the first "sensualists" (Englishmen, like Willis), who gave such an important role to sensation in the formation of ideas, as well as in the constitution of the intellectual faculties in general. The movement of the automaton was in no way independent of the will of the person who built it, who programmed it, and who set it running; just like Willis's "reflex," it obeyed a fate that was external—or, more precisely, superior—to it. Thus, a conception of movement was gradually established in which the acting subject was denied the possession of voluntary movement, in the sense of spontaneous or independent movement, and was subjected to the double determinism of the reflex and the rational soul.

The concept of the reflex, while still retaining the explanatory value that Willis gave it, evolved and gradually gained precision. It would not be until the work of J. Prochaska (published in Prague, 1779 and 1784), however, that the reflex would take on a definitive form: a movement or a behavior observed as a reaction to an excitation. For Prochaska, "fate" in the form of the motor response and the automaticity of its execution evoked the analogy of a beam of light on a mirror; it was well known that the reflected beam (movement) was equal to the incident beam (excitation). Nonetheless, Prochaska maintained, this reflection did not take place according to the laws of physics, but followed special laws "that we may know only by their effects, but may not understand." In fact, these laws (or at least certain ones of them) would not be explained until one hundred years later, first by E. Pflüger, later by C. S. Sherrington.

Prochaska's other major contribution was his theory concerning the site of the *sensorium commune*. He deemed it necessary to search at the origin of the nerves, where the sensory nerves arrive and the motor nerves leave, that is to say, in that part of the thalamus where the optic nerves terminate, in the brainstem where the other cranial nerves arise, and, of course, in the spinal cord. In any event, one did not look in the striatum and certainly not in the pineal gland. A reflex could thus arise from the "consensus" of a sensory nerve and a motor nerve, a consensus which could arise, according to Prochaska, outside of the *sensorium commune* proper, in subsidiary, decentralized centers such as the nervous ganglia.[4] Unlike Willis and a number of his own successors, Prochaska avoided constructing a generalized, explanatory system based on reflex movement. Preferring not to speculate about that which he did not know, he refused to accept the materialist thesis of the soul and thought, and contented himself with invoking the *vis nervosa* as a unifying concept, adhering to the vitalist thesis, just as the majority of the physiologists of his era did.

At the beginning of the nineteenth century, the attention of

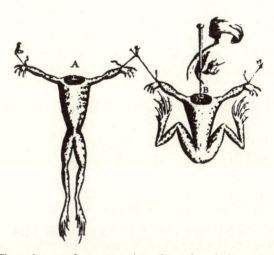

Figure 4 Experiment demonstrating the role of the spinal cord in the origin of movements. A decapitated frog remained immobile (A). Mechanical stimulation of the spinal cord produced a flexion of the posterior limbs (B). After destruction of the spinal cord, this phenomenon was no longer produced. This type of experiment, reproduced here by A. Stuart (1783), reported by R. Whytt, had also been performed by Leonardo da Vinci.

numerous physiologists turned to the spinal cord, which increasingly was being considered as a caudal extension of the brain and no longer as just a simple bundle that gathered together fibers from peripheral nerves. In 1750 Whytt had already reported experiments demonstrating that "involuntary" movement in a decapitated animal ceased when one destroyed the spinal cord stub (by penetrating the vertebral canal with a long needle). This observation attracted attention to the role of the spinal cord in the production of reflexes, and began to change the point of view of some workers, such as Prochaska and the French physiologists (Legallois, later Magendie and Flourens). By localizing the *sensorium commune* in the spinal cord, Prochaska gave it a role of its own and a certain anatomic dignity. The contribution of Legallois, based upon extensive experimentation, certainly represented a further step in this evolution. Having perfected a technique for artificial respiration, Legallois could keep animals alive while sectioning the brain or spinal cord by stages. Basically, if one were to "extract the brain by successive portions, from the front to the back, by cutting it in slices, one could in this manner remove all the brain proper, and subsequently the entire cerebellum and a part of the medulla oblongata."[5] It was at this stage that spontaneous respiration ceased and artificial respiration became necessary. As for movements, they persisted in a decapitated animal (which was already known) or in an animal with its spinal cord sectioned at the level of the first cervical vertebra, which amounted to the same thing, according to Legallois. The spinal cord was thus the source of sensation and "voluntary" movements,[6] contrary to the generally held opinion which placed this in the brain. Moreover, Legallois was able to isolate segments of the spinal cord from the rest of the central nervous system, again by transverse sectioning. He stated that "the parts [of the body] corresponding to each segment of the spinal cord experience sensation and voluntary movement, but without any harmony and in a manner as independent of each other as if one had transversely cut the entire body of the animal along the same planes; in a word, there were as many centers of sensation, each distinct, as there were segments of the spinal cord."[7] It was in the grey matter of the spinal cord that the "principle" which animated the different parts of the body resided, and it functioned by the intermediary of the spinal nerves. However, disconnected from the rest of the brain, segmentary movement was imperfect. Legallois thus formulated for the first time the idea of a "descending" control of the spinal cord by superior centers; if the spinal cord contained the "principle" of movements, the brain determined them, regulated them, and augmented their energy.

During the same period, the question of the direction of circulation of information in the nerves (the spinal nerves in particular) was resolved, an element essential to the understanding of the mechanisms of reflex movement. In 1822 Magendie experimentally demonstrated in the dog the polarity of the spinal roots: if the posterior roots were cut, the animal lost sensation in the corresponding cutaneous territory, while if the section involved the anterior root, it was the movements of the muscles involved which disappeared. The posterior root thus conducted in a centripetal direction: if one pinched it or pricked it, the animal evinced pain. It was therefore sensory. The anterior root conducted in a centrifugal direction; it was motor since, when it was stimulated, one observed strong (even convulsive) contractions of the muscles. The idea of a polarization of nerves was certainly in the air, and the Englishman Charles Bell had demonstrated an example the year previously, by distinguishing the sensory innervation of the face (by the trigeminal nerve) from its motor innervation (by the facial nerve). However, after the publication of Magendie's findings, Bell (seconded by his brother-in-law, a Mr. Shaw) became involved in a long polemic discussion, attempting to show that it was he, Bell, and not Magendie, who deserved credit for first discovering these findings about the spinal nerves. In point of fact, Bell had, in an unpublished paper in 1811, shown an intuitive understanding of the functional quality of the spinal roots. In his view, the spinal cord was divided into an anterior bundle which related to the cerebrum and a posterior bundle which related to the cerebellum. Thus, the anterior roots (which were connected with the cerebrum) could receive information coming from the periphery and send the volitional commands to the muscles, while the posterior roots (which were connected with the cerebellum) could direct the actions and functions of the viscera. In short, there was nothing really new in Bell's schema except an extension of Willis's concepts to the peripheral nerves; Bell's schema erroneously attributed sensory *and* motor functions to the anterior roots, and a "vital" function to the posterior roots.

It should be remembered, as A. Vulpian pointed out in his work of 1866,[8] that Bell experimented upon recently sacrificed animals, that is, on a preparation where his theoretical views could be partially confirmed but could not be disproved. Stimulation of the anterior root by a scalpel point produced muscular convulsions, which demonstrated the anterior roots' motor role. But neither the sensory manifestations induced by stimulation of these roots, nor the "vital" manifestations induced by stimulation of the posterior roots, could be observed since the animal was dead. In fact, by his experiments Bell certainly proved the nonmotoric nature of the posterior roots, but was only able to guess at their true function. It is because of all this controversy that today one speaks of "Bell and Magendie's law" to refer to the segregation of afferent and efferent messages entering and leaving the spinal cord.[9]

The study of the physiologic role of the spinal cord in the production of movements was marked by another personality, and

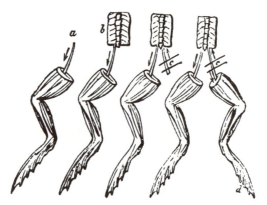

Figure 5 Experiments of Marshall Hall (1850) which demonstrated the reality of the reflex arc. Above, stimulation of nerve *a* of an isolated right frog leg produced contraction of this leg, just as stimulation of the cord (*b*) did. In addition, stimulation of the nerve to the left leg (*c*) or the left foot itself (*d*) could also produce contraction of the right leg. This demonstrated transmission of excitation through the cord. This experiment anticipated those of Sherrington fifty years later.

another quarrel, ten years later. In 1832 Marshall Hall consecrated more than 25,000 hours of his free time (by his own admission) to the study of the reflex. He came to the conclusion that the spinal nerves and the cord formed an independent system. Hall called it an "excitatory-motor" system, consisting of an excitatory limb, a spinal relay, and a motor limb. Reflex activity was produced by a "diastaltic" passage across the spinal cord, not through its "fibrous part" (which connected the brain with the body), but in its "central part," the "spinal marrow." Hall's "reflex arc" was unrelated to the will, consciousness, or sensation. The reflex was thus a separate category in his classification of movements, which consisted of voluntary movements controlled by the brain, respiratory movements produced by the brainstem, involuntary movements produced locally and, finally, reflexes of spinal origin. There was obviously nothing truly original in these conclusions, above all for someone who, like Hall, knew the work of Prochaska and Legallois. But who can actually claim to have produced a concept in this field which owes nothing to previous work? To the contrary, hypotheses in neurophysiology, perhaps more than in any other discipline, need to be constantly reformulated, enriched by new experiments or new speculations, in order to survive or to be definitively refuted. Still, Hall is criticized to this day for his failure to cite his sources, an omission that was to weigh heavily on his career. The same charges were not leveled against Müller, who "rediscovered" the importance of the reflex phenomenon around the same time as Hall.[10]

In his synthesis of this question in 1866, Vulpian pointed out that it was no longer a question of priority: everyone occupied the place due to him in the chronology of events, with his specific findings. The reality of the reflex was not only accepted as fact, but also as a means of "forming an idea" of the mechanism of action of the nervous centers. A reflex was thus defined as "a movement in a part of the body produced by an excitation coming from this part, and acting by the intermediary of a nervous center other than the cerebrum, and, as a result, without the intervention of the will." "Thus," continued Vulpian, "if one removes the head and viscera of a frog, and one pinches the foot, the limb withdraws. This is a type of reflex action produced by the spinal cord."[11] A similar mechanism was cited to explain blinking of the eyelids in response to a threat, sneezing, coughing, vomiting precipitated by appropriate stimuli, the contraction of the pupil in response to light, and also salivation and secretion of gastric juice precipitated by the introduction of food into the mouth or stomach.

A reflex is the simplest manifestation of the activity of the nervous centers, and at the same time a phenomenon that is, perhaps, not quite as simple as it at first seems. If one progressively increases the intensity of the stimulus applied to a decapitated frog's foot, one first notices a retraction of the stimulated leg, then a retraction of the opposite leg, and finally a contraction of all four legs. The excitation first spreads to the other side, then throughout the cord, by the intermediary of the grey matter. The very complexity of the organization of reflexes, which slowly revealed itself to experimenters, was encouraging to those looking for a function which exceeded the simple reflection of a stimulus. Moreover, it is possible that the abandoning of dualism as a physiologic explanation favored the formulation of a new teleology. There were those who were preoccupied with finding an explanation for behavior in the behavior itself, according to some law of causality for the individual (or even for the species), which expressed the fulfillment of some need or the achievement of some function. We have often used this term "function" in its contemporary sense, but it was really around 1870 that it took on this meaning: the individual, in order to adapt and survive, had to be able to produce appropriate responses to the situations he encountered; he did this by the exercise of some function, independent of volition or choice, since he did not know nature's intention. In 1877 Pflüger affirmed that nature is organized so that the cause of every need of a living being is also the cause of the fulfillment of that need.[12]

The concept of the internal milieu, which Claude Bernard described during this same period (1878), was also teleologic: "The constancy of the internal milieu is the condition of free, independent life." This form of life is "the exclusive domain of these beings which have arrived at the summit of complexity and organic differentiation." Because the most evolved animal is in a "tight and wise relationship" with the internal milieu, the variations of this milieu must be compensated and equilibrated at any given moment. The internal equilibrium of the organism "is a result of a continual, delicate compensation that is established, as by the most sensitive balance."[13] Bernard and Pflüger thus discovered homeostasis: the organism detects variations in the environment, and compensates for them in order to maintain its own functioning as a constant. It acts as a function of the goal itself, an order it must follow blindly. Didn't a reflex also have a homeostatic function, by producing "reactions which seemed to run in an intentional fashion to achieve a determined goal"?[14] This goal was "individual conservation," according to Prochaska. If the frog, even reduced to a posterior fragment, withdrew its foot, it was to remove the foot from a painful stimulus. If it contracted two feet, it was in order to jump and flee. For Pflüger, the spinal cord thus had a "psychic" power: after placing a drop of acid on the thigh of a decapitated frog, he noted that the animal, while flexing its limb, was able to rub the irritated area with its foot. If one them amputated the foot, the frog achieved the same end; after several fruitless tries with its mutilated foot, the frog eventually used the other, still intact foot. The spinal cord was thus quite capable of *perceiving* aggression, its localization, and its nature, and was able to elaborate the best response to combat it.

The apparent finality of the reflexes in the spinal animal led Pflüger to invoke a form of "consciousness" in the spinal cord, in the line of vitalist thinking. Consciousness (the soul) was coextensive with the nervous system, and was not only in the brain; it was divisible and could, as a result, exist in isolated segments. A lively controversy arose around this subject, due in part to R. Lotze, who categorically opposed this point of view. For him, if a spinal animal executed adaptive movements, it was because when the animal was intact the movements that occurred under the influence of the brain left behind them traces which allowed a stimulus to produce the same response as before, even when the brain had been removed. Therefore, it was not the effect of an intelligence that still existed, but the result of an intelligence that had formerly existed. For Lotze, it was clear that an animal which had never had any experiences

would be incapable of producing any "adapted" reflexes after ablation of its brain. Adaptive movements were voluntary movements
that had taken root as reflexes, by force of habit and repetition. A
spinal animal to whom one presented an unknown stimulus would
not respond.[15]

Such a controversy is, of course, endless, since knowing whether
or not the soul is divisible is a matter of belief, and not of objective
reality. The authors who grappled with this question after 1880
therefore restricted themselves to the observable facts: Was the
spinal animal capable of spontaneous or intentional movements? F.
L. Goltz, in Strasbourg, demonstrated that a spinal frog remained
immobile as long as one did not stimulate it, but it reacted in an
adaptive manner when stimulated. Vulpian had reported the same
thing: "A posterior fragment of a frog or a newt does not have the
least spontaneity. Volition . . . thus resides entirely in the anterior
fragment . . . It thus seems that it is the cephalad fragment which
alone possesses voluntary movements, since it moves without ap-
preciable external stimulus, while the pelvic fragment stays inert
as long as no external force stimulates it."[16]

The most credible example in this area, however, was Goltz's
series of long-term experiments in dogs, where he destroyed the
brain with a pressurized jet of water injected through a hole in the
skull. One such animal was able to live for eight months, as Goltz
showed at several scientific meetings. Unstimulated, the dog without a cerebrum seemed to be asleep. A loud noise woke it up, a
painful stimulus made it growl. Placed up on its legs, it would walk.
It was able to swallow food placed in the back of its mouth. These
results were interpreted by Pflüger's defenders as proof that the
spinal cord could "feel" and organize action in response to sensations
which it received. Was it reasonable to proclaim, as G. H. Lewes
did in 1873, that all actions occurred via this mechanism, and that
which was called "spontaneous" was due, in fact, to an internal,
unobservable stimulus? If a dog wagged its tail while dreaming, or
when it saw its master, one would say that this was a spontaneous,
voluntary movement. If the dog without a brain wagged its tail,
one would say it was a reflex. Why this distinction? asked Lewes.
While not rejecting the distinction between reflex and voluntary

movement, he refused to consider a reflex as the result of a purely "mechanical" process. For Lewes, the property of any action was "sensitivity," a vital property which distinguished action from any physical or chemical process. In the extreme, then, all actions were reflexes to the extent that they all obeyed the same vital principle, being precipitated by a sensation.

3

Hierarchy and Integration of Movements

$\mathcal{T}$he development of the concept of the reflex changed underlying assumptions about the classification of movements. Could one still take for granted the distinction between voluntary and involuntary, when even the most complex behaviors might obey the same laws as the simplest reflexes? "The terms 'reflex', 'involuntary', 'voluntary', 'automatic' are more than classificatory designations; they have come to carry a burden of implications, philosophical, physiological and psychological as to the nature of nervous function. It is essential that they be used with caution, and that the hypothetical implications which they have acquired during the seventeenth and eighteenth centuries be regarded as provisional only."

This warning of F. Fearing is not superfluous.[1] If a border did, in fact, exist between voluntary and involuntary movements, it was not easily drawn. It was certainly not an anatomic one, and it evaded the physiologists. Might it be accessible to philosophers only?

Maurice Merleau-Ponty's conception of "reflex behavior" is interesting in that respect, because it went beyond the classical duality between reflex and voluntary movement. In the first place, he said, the reflex exists, its cause is a stimulus, and the organism which produces it is passive, "since it limits itself to the execution of that which it is instructed to do, by the location of the excitation and the nervous circuits which take their origin there." It is "an apparatus mounted by nature."[2] But the reflex apparatus was not an isolated apparatus; it depended upon the neural and motor situation of the organism as a whole. The brain acted upon it, not by giving or refusing the authorization to function but by taking part in the constitution of this nervous and motor state. The brain acted by making "one mode of organization predominate over another,"[3] in the extreme case by making voluntary behavior predominate over reflex behavior. Behavior could not be envisaged in terms of separable elements, but in terms of "levels of activity," which represented varying degrees of adaption to the environment. "It would conform better to the facts," wrote Merleau-Ponty, "to consider the central nervous system as a place where an overall 'image' of the organism is elaborated, where the local state of each part finds itself expressed . . . It is this image of the whole that commands the distribution of motor influx, and gives the impression of organization to which the least of our gestures testifies."[4] The nervous system is the generator of order and coherence, a biologic and human order, as we shall see. Its function harbors a "principle which *constitutes* order, rather than being subject to it."[5]

The modern conception, which integrates the reflex into the totality of the nervous system, and thence into the functioning of the organism, is the product of a long evolution in two, apparently opposite, directions. The controversy between Pflüger and Lotze contained the essences of these two tendencies. Pflüger supported the notion that the most complex motor behavior was only the summation of simple reflexes; Lotze supported the thesis that a simple reflex was only an arbitrary unit of behavior, which always remained under the control of a higher "center." Even though this controversy was one of those quasi-ideologic debates, where the ephemeral victory of one party over the other rests more on the coherence of the theoretical construction than on the experimental facts which support it, this conflict nonetheless weighed heavily on the orientation of research.

The first position, toward which a great number of the physiologists of the second half of the nineteenth and the beginning of the twentieth century tended, was associated with the philosopher Herbert Spencer, who himself had been strongly influenced by experimental work on the reflex. During this period, it was common practice to take physiologic findings and interpret them in a certain philosophical framework before passing them on to the following generation as a new concept of behavior. In his work of 1855, Spencer appeared to hold with the construction of "psychism" and, at the same time, to promote an evolutionist theory (several years prior to Darwin).[6] He had a double thesis: the reflex was already a psychic act, and the psyche was an assemblage of reflexes; instincts, formed from reflexes frequently repeated, became fixed and were transmitted in a hereditary manner to following generations. These ideas, adopted in a modified form by Darwin in *The Descent of Man* (1871),[7] made behavior and its cerebral determinism subject to evolutionary pressure, in the same way that morphologic characteristics were; for Darwin, the movements of the face expressing the emotions, for example, were innate and hereditary, and appeared in the course of evolution in an "independent and natural" manner. Human behavior, with its apparent complexity and perfection, was therefore separated from animal behavior only by a matter of degree, not nature, and species evolved from the simple toward the complex, from the inferior toward the superior.

A nearly direct application of Spencer's ideas was found in the theories of Hughlings Jackson regarding the neural organization of movement. Jackson had been strongly influenced by his teacher, the neurologist T. Laycock, himself an intellectual disciple of Spencer. Laycock professed that the brain, the organ of consciousness, was subject to the laws of reflex action like any other neural relay; the brain constituted the central link of a reflex, where an "ideogenic" process intervened. He argued that the sensory experience of an object, or even the simple idea of this object, could evoke the ideogenic process and produce the same motor response. Jackson adopted a classification of movements that was based on their degree of automaticity. The same muscles—for example, the respiratory muscles of the thoracic cage—could participate in completely automatic movements (respiration), slightly less automatic movements (maintenance of posture), or nearly nonautomatic movements (speech).

These different categories, from the most automatic to the most voluntary, reflected an evolution of the neural centers from the simple to the complex, that is, from inferior centers responsible for purely automatic movements to superior centers responsible for minimally automatic movements. It should be noted that the Spencerian term "evolution," which Jackson used, could be taken as easily in its phylogenetic as in its ontogenetic sense. This contained the notion that the more primitive animal species had neural centers inferior to those of more evolved species, and as a result they were only capable of automatic movements; it also contained the notion that the development of the nervous system in a given individual passed through increasing stages of complexity. Jackson therefore thought that the inferior centers were organized at birth, while the superior centers developed later in life. The simplest centers had a rigid organization, while the most complex ones had, by contrast, a looser organization. At the apex of evolution, the centers where the mind had its origin, and where the physical basis of consciousness resided, were the least organized, most complex, and most voluntary of all.

Let us not forget that Jackson was a neurologist. His theory of the evolution of neural control of movement was, in fact, only the inverse of an explanation of neurologic illness, and a rationale of neurologic symptoms. Neurologic illness was, in effect, a dissolution (another term of Spencer's) and a regression of evolution, a reverse process that led behavior back to its most automatic, simplest, most rigidly organized forms. A destruction of the superior motor center, in the cerebral cortex of one of the hemispheres, produced a paralysis in the limbs of the opposite side, a negative symptom expressing dissolution; this simultaneously produced an exaggeration of certain spinal reflexes, a positive symptom signifying regression. The dissolution of the superior level of behavior liberated the inferior level; the extent of the dissolution determined the nature of the symptom, "neurologic" if the dissolution were local and "psychiatric" if it involved the entirety of a given level and if it were uniform. In the latter case, the involvement of the "organ of mind," the superior level of the structural and functional hierarchy, was not only responsible for disorders in the consciousness of self and dementia—both negative symptoms—but also for hallucinations, delirium, and automatism, which were positive symptoms expressing the activity of inferior centers. Jackson's theory naturally had a profound influence upon psychiatry. In France in particular, a school of thought (the organo-dynamism of Henry Ey) was explicitly based on it, proposing that it is not the manifestation of the unconscious which explains psychopathologic symptoms but rather the modalities of disorganization of the conscious being.[8]

Stratification or hierarchy? Did Jackson conceive of the organization of the nervous system as the superposition of autonomous reflexes, the cerebral part of the edifice functioning according to the same model as the medullary part, or as a hierarchy of interdependent levels, the highest level controlling (and, on occasion, repressing) the subjacent levels? The notion of liberating inferior levels by "dissolution" seemed to imply an integration or subordination of these levels to the nervous system as a whole. According to Henry Ey, the Jacksonian hierarchy of neural levels and functions implied a teleologic point of view: the hierarchy became the "plan," the "logic" of the system, with inferior or instrumental functions subordinating themselves to superior functions, "just as words to syntax, means to the end."[9] In other words, departing from Spencer and his mechanistic model of reflex organization, Jackson progressively moved toward a more supple model, which implied the interdependence of the parts and their subordination to the whole. One can sense this evolution in thinking when we examine Jackson's discussion of the relationship, at the top of the functional hierarchy, between the mind and its neural organ. The brain is the instrument of the mind but does not constitute it; the mind and the brain are concomitant, parallel. This neodualist attitude, also adopted by Sherrington, was common in England at the end of the nineteenth century.

However, Spencer had disciples more faithful than Jackson. Ivan Setchenov, the father of Russian physiology and psychology (and thus the spiritual father of Pavlov), was among them. While experimenting in Bernard's laboratory in Paris, Setchenov noted that the foot-withdrawal reflex in the frog was diminished when a crystal of table salt was applied to the animal's brainstem. Inversely, the same reflex was augmented when the brainstem was transected.

Thus, one might conclude that the brain could exercise an inhibitory influence on the spinal reflexes. Setchenov had, in fact, just discovered inhibition in the central nervous system,[10] a concept now inseparable from excitation. Setchenov, however, thought that the brain was itself a reflex center, where all the responses of an organism were elaborated. Sensory afferents arrived there, and muscular efferents left there; during the course of this central passage, augmentation (excitation) or diminution (inhibition) could be produced from the basic reflex response. For Setchenov, the intensity of the stimulation determined the trajectory of excitation in the interior of the nervous system: "If a man is subjected to the action of an external agent, and is not frightened by it, then the intensity of the ensuing reaction (whatever muscular movement it may be) will conform to the intensity of the external influence." This is the very definition of the reflex according to Prochaska. However, if the stimulation provoked fright, the motor reaction of the subject became violent: in this case, "the new mechanism intensifying the reaction is added to the activity of the old mechanisms which produce the reaction."[11] The influence of Spencer is obvious, all the more so since Setchenov had this reflex apparatus play an adaptive role, guided by "the instinct of self-conservation." The simple foot withdrawal reflex in a decapitated frog was by nature no different from a complex flight behavior: the apparent rationality of a movement did not exclude the mechanical nature of its origin.

Were all "involuntary" movements independent of mental activity and thought? This was doubtful, since they protected the organism in a rational manner and, ultimately, they seemed caused by motivation that could be considered "psychic." At this point in his treatise (in the chapter on involuntary movement in his 1863 book) Setchenov was not, fundamentally, very far from Jackson. Moreover, the notions of the teleologic organization of the simple reflex, and the psychic nature of involuntary behavior, were present in varying degrees among most of the authors of the second half of the nineteenth century. The intensification of a reflex movement by cerebral centers higher in the hierarchy is in line with the concept of integration, already present in Jackson's work and to become the center of Sherrington's work. Finally, even the conclusion of Setchenov's first chapter, "all involuntary movements are mechanical in origin,"[12] seemed to leave room for another category which defined itself by opposition: voluntary, nonmechanical movements. In fact, Setchenov radicalized his thinking in the second part of his book, and justified the title which he had first given to it. In an autobiographical note, he neatly explained his conduct: "All the movements known in physiology as voluntary movements are reflex movements in the strict sense of the word." The title initially proposed for the work, *An Attempt to Establish the Physiologic Basis of Psychical Processes*, was rejected by the tsarist censorship; Setchenov changed it (perhaps ironically) to *Reflexes of the Brain*. This, he said, "earned for me the charge that I was an involuntary propagator of immorality and nihilist philosophy." "There was no need to discuss in my treatise the matters of good and evil; my object was to analyze actions in general, and my point was that given certain conditions, not only actions but also their inhibition are inevitable, and obey the law of cause and effect. Is this an apology for immorality?"[13]

How does one begin a physiologic study of voluntary movement? By admitting that one knows nothing about it, and by basing the work only on solidly established facts. But this "naturalist" attitude of Setchenov rapidly came into conflict with his project, or his desire, to interpret voluntary movements as the action of a relatively simple mechanism: "Thus we have to prove . . . that this particular human activity consists of reflexes which begin with sensory stimulation, continue in the form of a defined psychical act, and end in muscular movement."[14] That which distinguishes voluntary from involuntary movements is only the fact that voluntary movements may serve higher motives, such as the welfare of mankind or the love of country, and often contradict the instinct of self-conservation. Apart from this difference, all the elements of the reflex are present in the voluntary movement. If the external cause seems to be missing, it is because it is imperceptible; but the psychologist who follows the rule of objectivity must attempt to discover the initial excitation, the key to understanding all behavior. Sounds, images, and sensations could all be preserved in a latent state in the nervous system between the time they are, in effect, "obtained" and the time they are "reproduced," thus playing the role of stimulus for an action which merely appears to be spontaneous. Setchenov thought that nerves could conserve traces of an excitation for a long period afterward. This nerve "memory" was not a mental construct and was not of a psychic essence: it could be explained entirely by the physical laws of repartition of an excitation into the interior of different fibers of the same nerve. "There is not the slightest difference between an actual impression . . . and the memory of this impression."[15] Moreover, in the same manner that the cause of a reflex (its precipitating stimulus) could pass unnoticed by an external observer, the response itself could be absent. It would merely be inhibited. The psychic reflex was thus reduced to a purely internal event, a thought, a mental image, an evocation "private" in appearance, which was in fact precipitated by an imperceptible external event, and whose expression had been blocked.

"The initial cause of any human activity lies outside man."[16] This is the core of Setchenov's thought, influenced, as he recognized, by the French sensualists of the eighteenth century. To those who spoke to him of "desires" and "wishes," he answered: You are only following your reflexes. Setchenov had an imaginary reader speak thus: "I think that I wish to bend my finger in one minute; I actually do so" (the reader does, in fact, bend a finger at the end of one minute); "I am absolutely conscious of the fact that the beginning of this act originates within myself."[17] This is only an illusion: this act in its entirety is only a series of "associated thoughts" evoked by conversation and expressed by a movement. The ideal of total objectivity seemed to be attained, since the external observer who knew all the external stimulations received in each instant by an organism, and who recorded all the movements of that organism, would know *ipso facto* its mental state. The psyche was thus explained, and voluntary movement found itself once more reduced to a matter of style, to a category without object, to a level in the hierarchy of reflexes.

The concept of *brain reflexes* constituted the point of departure for the work of I. P. Pavlov and his school, at least to that part of his work which was devoted to psychophysiology. Pavlov, who received the Nobel prize in 1904 for his work on the digestive glands, published in 1897, was observing animals with chronic gastric fistulae when he noticed the "psychic" secretion of the stomach. In 1903 he wrote: "What importance is there in considering these phenomena "psychic," as opposed to simple physiologic phenomena, since we know that the naturalist cannot distinguish between them except from an objective point of view, without concerning himself with their essence."[18] Vitalism confused the points of view of the naturalist and the philosopher: to the philosopher belonged the attempt to "create a whole out of the objective and the subjective," while to the naturalist belonged the discovery of the truly objective method. As Pavlov's own path of research showed, the two could only be one when the objective method of study was applied to the superior functions and to subjectivity itself. Pavlov "bowed" before the work of the psychologists, but felt their "output" was bad, remaining "convinced that the pure physiology of the animal brain will lighten . . . the work of those who devote themselves to the study of the subjective states of man."[19]

But what is "the pure physiology of the brain"? Abandoning the terms "involuntary" and "voluntary," which Setchenov had kept, Pavlov felt that the relationship between the organism and its surrounding environment was ruled by two types of reflexes. Certain of these reflexes were permanent, innate, and characteristic of the species: these were "absolute reflexes." Others were "conditioned reflexes." These new reflexes appeared "each time an indifferent agent coincides with the action of an absolute excitation of the reflex." "This agent becomes, after one or two coincidences, capable of itself provoking this same reflex act."[20] It is thus that the nervous system became "creator," and no longer merely "effector," in the sense that it was ceaselessly creating new connections between the organism and the external environment. The evolutionary viewpoint was certainly not absent in Pavlov's thinking, as when he stated that it was "highly probable . . . that the conditioned reflexes acquired by an individual will become, little by little, absolute, when several generations have passed under the same living conditions. This mechanism would thus be one of the factors in the continual development of animal organisms."[21] Following this line of reasoning, one could understand how a species adapted itself to its environment, by sensitizing itself to the most infinitesimal natural phenomena; initially without any influence on the organism, these phenomena would end up becoming intensely excitatory for the vital functions of the organism. The conditioned reflex would thus belong to that fundamental tendency of all living things (and particularly of superior beings) to adapt to and place themselves in equilibrium with the surrounding environment. This would also be true for voluntary movement, achieved at the level of the "motor analyzer" of the cortex, by the association of stimulation arising from muscles and joints with those arising from some other origin (alimentary, for example); voluntary movement would thus represent a superior mode of adaptation.

In denying the autonomy of psychic or mental processes, that is, their spontaneity and their independence from the surrounding world, Pavlov belonged to the sensualist and empiricist tradition. But by admitting the existence of these same psychic processes, for which he elaborated the "materialist" thesis of conditioned reflexes, he distinguished himself clearly from the other objectivists of the era, Watson and the behaviorists, for whom the behavioral responses were the only bases of psychology, and not the revealed expression of an alleged internal state.[22] One of the central notions of the Pavlovian model is that these high-level reflexes which constitute psychic activity depend upon the cerebral cortex. The ablation of the cortex (decortication) would prohibit the acquisition of conditioned reflexes, and would suppress those which had been acquired before the procedure. The cortex contained the analyzers, "the neural apparatus whose role is to decompose the complexity of the external world into its elements." The destruction of the "cutaneous analyzer" in the cortex of a dog profoundly modified its behavior in response to tactile stimulation. Petting, which "in a normal animal would bring about a friendly reaction, in this animal, who has lost the superior part of its cutaneous analyzer, only brings about a reflex of an inferior type, arising from the inferior part of the brain, and manifesting itself as a defensive reaction of the animal."[23] The equation: higher nervous centers = superior behaviors had already been generally accepted before Pavlov, but the equation, *higher nervous centers = cerebral cortex*, would appear, following his work, as a veritable dogma, in the form of the aphorism, "No conditioned reflexes without cortex." Such a clear-cut statement, connecting mental activity with conditioned reflexes and imposing a cortical site on the whole, was exposed to a rapid refutation. How could one explain the capacity of species such as fish (who had no cortex) to acquire conditioned reflexes? How to explain the demonstration of phenomena such as habituation (which Pavlov had not hesitated to categorize as psychic) in insects or molluscs?[24]

The Spencerian interpretation of the role of the reflex as the elementary unit of psychic function, as the "atom of the psyche," could only lead to an impasse; thus, Pavlov, under the pretext of objectivity, closed the door to a true study of the internal state and the mental contents. Sherrington's conception, on the other hand, in many respects contradicted the concept of conditioned reflexes, but at least it preserved the possibility of examining the inner state. Sherrington, whose essential discoveries on reflexes were made between 1895 and 1900, after he had been named Professor at Liverpool, began with premises similar to those of Setchenov and Pavlov, since he also considered the "simple reflex" as the basic unit of neural function. His views differed radically from the Russian school, however, in the definition of this function. Sherrington did not seek to locate the role of the nervous system in some adaptation of the organism to its external environment, the Spencerian thesis *par excellence*. To the contrary, this role must be "within," as a force for ordering and organizing behavior. Sherrington's conception did, in fact, maintain that it was the nervous system which imposed its order on the environment, and not the inverse. One can thus understand the vehemence of his discussions with Pavlov, who had forbidden his students to read Sherrington's work. Sherrington, albeit less intolerant than Pavlov, nonetheless testified to the irreconcilable nature of the disagreement; and, when he visited Pavlov's laboratory, told him: "Your conditioned reflexes have little chance of being believed in England, because they have a materialist flavor."[25] In fact, E. G. T. Lidell, one of Sherrington's own students, did not even cite conditioned reflexes in his book regarding the discovery of reflexes—written in 1960.[26]

For Sherrington, the reflex arc was the basic element of the nervous system, but only when one considered the system in its *integrative* function; the reflex is an elementary reaction, but it is the elementary reaction of neural integration. On the other hand, taken as such, the simple reflex "is probably a purely abstract conception, because all parts of the nervous system are connected together, and no part is probably ever capable of reaction without

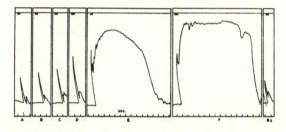

Figure 6 The method for recording movement was based for a long time upon the principle of the kymograph, first developed by J. Marey. The animal preparation was fixed on a platform. The extremity (or muscle) whose movement was being observed was attached to a lever. Displacements of the lever were recorded on a rotating cylinder, which was covered with lampblack.

Using this technique, Sherrington recorded the various aspects of reflex activity. A crossed extensor reflex (extension of the foot contralateral to the stimulated foot) is shown in this case. Each tracing (*A, B, C*, etc.) represents a response to an electrical stimulation, whose duration is represented in the form of a black rectangle in the upper part of the recording. Each division on the bottom of the tracing represents one second. Recording *A* represents the response to a stimulation of low intensity. Regular augmentation of the current intensity increased the response proportionally (*B, C, D*) up to the point where the response increased considerably, out of proportion with the stimulation (*E, F*). Phenomena of summation and diffusion of excitation were produced, "recruiting" new neural circuits.

45

affecting and being affected by various other parts."[27] The simple reflex was thus always integrated into an action which exceeded it, and which included it as a constituent element. An example of this is "reciprocal innervation": when an extremity is flexed, the extensor muscles are inhibited, while during an extension the flexor muscles are inhibited. The "simple" reflexes of flexion or extension thus necessarily include an inhibitory component, brought into play at the same time as the excitatory component. Actions such as walking or scratching are the result of the coordination of simple excitatory and inhibitory reflexes, simultaneously or in series. Even at the most peripheral level, the same muscles could be implicated in different reflexes, even reflexes of opposite "sign." The spinal cord motoneuron (the nerve cell which directly commands the contraction of a muscle) was like the keys of a piano, on which was played what might be called the "motor melody."

The motoneuron was thus the "final common pathway" where motor commands ended. An anatomic hierarchy (from the brain to the reflex, and no longer from the simple reflex to the brain reflex) controlled the spinal reflex, and directed the excitation to appropriate "keys" on the "piano." Interneurons—a new concept—were included in the intraspinal circuits, and achieved different modes of coordination of movements, such as irradiation (allowing the withdrawal of not only the stimulated segment of the extremity, but the extremity as a whole), reciprocal inhibition, or chaining (allowing complex actions to occur). These interneurons served a role analogous to rigging on a sailing vessel, distributing the excitation and inhibition in one direction or another as a function of the action to be executed. But what presided over these actions? What selected their goal? What knew their probable effect? In other words, where was the pianist who played the keyboard?

Legallois, in 1812, had given one response to these questions by placing in the brain the authority "which determined and ordered all the acts of animal function," and offered as proof the fact that the movements of a decapitated frog were "disordered, and without purpose."[28] Sherrington, nearly a century later, took up this classic thesis and renewed it. In effect, the movements produced by a spinal animal (spinalization, or high sectioning of the cord, as we have seen, is the neurologic equivalent of decapitation) are not

random and meaningless, but they are "mindless." These movements can be coordinated (as in withdrawal or scratching), but they
only represent an impoverished image of normal behavior. The
"mindless body" is merely a machine that reacts to physical stimuli,
but no longer reacts to psychic excitation. This dualism was even
more openly expressed in Sherrington's later writing, for example
in 1933 when he stated that he could find no correlation between
mental experience and cerebral events: the two could, undoubtedly,
coincide in time and place, "but this is of no help since . . . neither
of the two appears in relation to the other."[29]

Although Sherrington made his dualism a matter of personal
conviction, it eventually gave way to a rational explanation of the
mechanism of movement. The explanation centered around the
terms "coordination" and "integration." The simple reflex was integrated into an action directed toward the achievement of a goal;
this goal coordinated the different motor elements which eventually
led to its achievement. The brain held under its domination the
subjacent levels, in particular the spinal level. Could one complete
the reasoning by placing the goal "in" the brain? This was the
question, posed in similar terms (structurally speaking) by Jackson,
but including an additional degree of complexity. Jackson had, in
fact, envisaged a hierarchy and an anatomic integration of behavior,
but had not posed the question of their submission to a biologic
order determined by the brain.

The notion of integration into the whole led certain schools
(particularly the German school of the 1920s and 1930s) to a complete refutation of the theory of the organization of behavior starting
with the reflex. Thus, the explanation of behavior moved back and
forth from Flourens onward, like the swinging of a pendulum; in
turn, the partisans of an exclusive role for environment and the
partisans of a role no less exclusive for the organism seemed to hold
sway. Kurt Goldstein, one of the most representative authors of
the latter school, stated that as far as he was concerned, reflexes
were not organized in a rigid fashion. The force of the response,
the dimensions of the cutaneous area which the excitation stimulated, the form of the response itself, varied as a function of the
total state of the organism. One of the best examples of this dependence, according to Goldstein, was the inversion of the cuta-

neous plantar reflex in man following a lesion of the pyramidal tract. Stimulation of the plantar surface of the foot normally produces flexion of the great toe, but an extensor response (Babinski's sign) is noted under pathologic conditions. Goldstein arrived at the conclusion that "nowhere can one observe directly a type of relationship [between stimulus and response] which corresponds to the strict concept of a reflex."[30] To obtain a reflex, it would be necessary to completely isolate the reflex arc from the rest of the organism, which is only possible in an artificial manner; to explain the modification of the reflexes according to the concepts of Sherrington (facilitation, irradiation, etc.), it would be necessary to usurp the right to consider such a phenomenon as a "normal" reflex, which was arbitrary. Thus, it was not the environment which created order for the organism. At the extreme, thought Goldstein, one might say "that an organism cannot exist unless it succeeded in finding in the world . . . an adequate environment." "An environment only comes about in the world for a given organism. Thus, order must receive its determination elsewhere. Where? From the organism itself."[31] Goldstein derived his certainty from observations of humans with cerebral lesions: "This altered organism, for whom the usual environment has become strange and troubling, can only exist under the fundamental condition of being able to extract from the world a new adequate environment."[32]

This extreme hypothesis rendered the organism or its nervous system an entelechy, whose only reason to exist was to assure its own survival. On the experimental plane, it coincided with an ingenious series of experiments intended to test the degree of "specificity" of the reflexes and to demonstrate that the relationship between stimulus and response was purely contingent upon the momentary relationship between the organism and the environment. As early as 1827, Flourens had the idea of transposing the nerves normally responsible for raising and lowering the wing in a chicken. The animal with lowering muscles innervated by the raising nerve, and vice versa, could fly as well as before the experiment, according to Flourens. This same type of experiment was subsequently performed in numerous species by different researchers. Apart from a few discordant voices, the general opinion was that the new innervation fulfilled, without fail, its new function: the direction of voluntary motor impulses seemed to be able to be modified at will. "The nervous centers, by training, can furnish exactly what the peripheral organs with which one has connected them require."[33] Training or exercise were not even necessary, according to Goldstein, and correct innervation was evident immediately after the operation. It is therefore not surprising that this type of intervention was used in humans subsequent to paralyzing trauma or an illness such as poliomyelitis. Certain postures—for example, extension of the hand or elevation of the foot—are essential for the preservation of prehension or normal gait. When one of the extensor muscles responsible for these postures is paralyzed, it can be replaced by transposition of another, less important muscle onto its tendon—a muscle which might have had a very different "function," for example, flexion. Since the transposed muscle keeps its original innervation, the neurons that originally commanded it to contract would recoordinate their activity. How was this possible, if the circuits which ended at the motor neuron were "wired" in a rigid manner? Moreover, it seemed that those subjects who had such muscular transpositions were able to achieve the appropriate use of their extremity when they concentrated on the task to be accomplished, but if they became distracted or tired the "old" coordination could transiently reappear.

Paul Weiss, who studied several cases of this sort, suggested in 1941 that the recoordination of the extremity's activity after such muscular reversal must be produced "higher," in the cortex, and not at the level of the cord. We should note the progressive shift: from the moment when the notion of reflex was placed in doubt, or at least from the moment when its deterministic and rigid role in commanding and executing an action was reduced, the complementary notion of voluntary movement resurfaced. In the case we have before us, Weiss felt that "new types of action" could be defined by the nervous system, without taking into account the arrangement or rearrangement of its connections with muscles, but only taking into account the goal to be attained and the "optimum product" (Goldstein's term) of the execution of this goal. In effect, for Goldstein, "it is quite immaterial that the new connection . . . be made by one pathway or by another. It suffices that the connection exists for innervation to be produced." The move-

49

ment that is intended to be accomplished is not linked to the excitation of a determined structure, but rather it represents a "figure defined by an excitation" which can use "any available structure" for its action.[34]

Goldstein was strongly influenced here by the "theory of resonance" conceived by Weiss. According to this theory, the excitatory center for a movement and the corresponding effector muscle were not linked by any particular connection: they functioned, one might say, on the same wavelength. It is sufficient for the center to emit the excitatory signal in order for the corresponding muscle, and this muscle alone, to recognize it. It matters little what pathway the signal uses to arrive at the muscle. Weiss, it should be noted, was interested above all else in the formation of neural connections in the course of embryogenesis. His theory, in postulating a random establishment of these connections, offered an alternative to that of rigid wiring which was executed in a "programmed" manner in the course of development. This hypothesis, on the other hand, set the stage for an organizing center of movement, functioning in an autonomous manner and, at the same time, imposing its order on the organism as a whole.

The emphasis placed on the goal to be achieved, and on the function to be assured, as the determiners (rather than the consequences) of biologic order brings us closer to voluntary movement as it brings us away from the reflex. Even if it represents undeniable progress in the understanding of action and movement, this attitude is nonetheless not without risk. The dualism of Sherrington and his more recent successors serve to remind us of this.[35]

4

The Motor Cortex

*T*he cerebral cortex is particularly attractive for physiologists and neurologists. Even early on, when its real function (or at least that which we attribute to it today) was still misunderstood, it was already viewed with a certain veneration, this folded mantel which covered the brain, whose expanse grew with the evolution of species. For Willis, it was the cortex which distilled the animal spirits, and in its folds that the images born of sensations became lost. Thanks to the anatomists at the dawn of the nineteenth century, and particularly to Gall, it became a differentiated structure, whose elements were recognizable and accessible to description. By tracing the fibers of the spinal cord and the medulla with a scalpel, Gall determined that their tracts continued right up to the cortex. Origin or termination, this was the highest stage of the nervous system. Its anatomic distinction, combined with the enfolding which permitted an increase in surface area without notable increase in volume, conferred on the cortex, and above all on the human cortex (the most developed of all), a privileged role in the explanation of behavior. Even in the Pavlovian model, where its dimensions are reduced to a reflex center, the cortex maintained its place apart from the rest: it was here that the analyzers of sensations were found, and here that new connections, created from the association of stimulations, were made. It is interesting to note that all the theories of cerebral organization that we have reviewed so far postulate the existence of a "superior level," which finally found its materialization in the cerebral cortex. From the end of the eighteenth century onward, it was the cortex that played the role of that mysterious but necessary entity which one no longer dared place outside the brain. Receptacle or even, in the extreme, tabernacle, only the cortex could henceforth be the *ultima ratio* of neural activity, intelligence, and will.

In the domain of movement, the study of cortical mechanisms prior to 1870 was limited to physicians' and physiologists' observations of the effects of lesions in animals and man. Pourfour du Petit, physician in the King's Hospitals, is usually credited with the discovery of the crossing of motor pathways. Observing those war-wounded who were afflicted with paralyses of an extremity or one side of the body, he noted that the lesion was located at the level of the cortex (not the ventricle) on the side opposite the paralysis. Thus, he deduced quite correctly that the animal spirits which were filtered by the right side of the brain served to move the left side of the body, and vice versa. The clinicoanatomic method of comparing symptoms with the location of the lesion (generally revealed by autopsy) rapidly bore fruit. In 1762 Joseph Baader wrote that "the elements and the action of the brain are subject to decussation, so that the sensation and motility of one side of the body are under the dependence of the opposite cerebral hemisphere." In reporting a clinical observation, he noted that "our patient complained of the right side of his head, and on this side an abscess was found, while . . . the convulsions always involved the left arm." It thus became possible, according to Baader, to "know and predict, for the greatest benefit of practitioners, what part of the brain gave this or that extremity sensation or movement; as a result, knowing which member is involved, one can determine what part of the brain is ill, and inversely, being given a certain lesion in the brain, predict which member must be affected." Having observed another similar patient, Baader suggested that "the region of the brain which is located beneath the parietal bone commands the motility and the sensitivity of the upper extremity of the opposite side."[1] The pyramidal tracts, which connect the cortex with the spinal cord and cross at the base of the brainstem, according to Turck's description of 1851, could apparently explain these crossed effects.

For Bouillaud and the neurologists of 1825, the cortical origin of motor troubles such as convulsions or paralysis, whether a hemiplegia (paralysis of the half of the body opposite the site of the lesion) or a monoplegia (paralysis of a single extremity), was no longer in question. Beyond the routine neurologic examination, an attempt was being made to localize in a precise manner the cortical movement "centers." In 1778, thirty years before Gall, Sabouraut had already thought that if one could follow the nervous fibers from a paralyzed muscle to their origin in the brain, one might determine "at what location in the brain the disorder was . . . A lesion of this location in the brain must necessarily cause some particular involvement of the functions of those parts of the body where these nerves end."[2] For Bouillaud, who knew and admired the teachings of Gall, it was possible to delimit the "cerebral organs devoted to movement," of which the majority "were proved merely by the existence of partial paralyses corresponding to a local alteration of the brain."[3] One part of the cortex might control movements of the arm, another those of the leg, or tongue, etc. In the general conclusion of his text of 1825, Bouillaud noted: "1°) in man, the brain plays an essential role in the mechanism of a great number of movements; it controls all those which are subjected to the empire of intelligence and the will; 2°) the brain has several special organs, each one of which has, under its dependence, particular muscle movements."[4] Contrary to what Flourens thought, the will, and likewise the intelligence, were not properties of the cortex as a whole. The will to move one or another extremity was quite localized in a certain part of this organ.

How could this point of view be reconciled with Flourens's experimental results, which tended to show that an animal whose cortex was removed lost all spontaneity but continued to move normally if the observer stimulated it? Was what Flourens taught, as Bouillaud thought, "experimentally inexact"?[5] Let us rather say that Flourens, influenced by his own reticence at localizing limited functions in the cortex, placed himself in an experimental situation where he could only obtain the results he wanted. Bouillaud noted that a man with a hemiplegia resulting from a localized cortical insult, in whom the voluntary movements of one side of the body were abolished, nonetheless had preserved "automatic and instinctive" movements: the patient withdrew his leg when pinched. Moreover, in an animal experiment Bouillaud demonstrated that the ablation of the anterior part of the cortex produced a state of "idiocy" in the dog, wherein the animal lost its ability to seize and recognize the relationships between things, despite the preservation of normal vision. This faculty of intelligence was thus distinct from the sensory faculties which, according to Bouillaud, were localized in the posterior part of the cortex. The reasoning was, by and large, correct, even if Bouillaud did not perform the experiment which would have been able to demonstrate this with certainty. He was satisfied with the partial experimental confirmation of his clinical observations of paralyses and language troubles. Being able to localize just a single function of the cortex allowed one to envisage the localization of all the rest.

The localization of cortical areas implicated in movement was certainly a crucial step, even if, as we will see, it did not resolve the theoretical problem of the relationships between cortex and voluntary movement, the object of a discussion which is still open. It is to E. Hitzig and his associate G. Fritsch that we owe the discovery, in 1870, that certain parts of the cerebral cortex are "excitable." Attempts at applying electric current to the surface of the brain had, until then, given negative results, from whence came the dogma of the inexcitability of the cerebral substance. Since the work of Fritsch and Hitzig, however, electrical stimulation has become the complement to lesions as the way to demonstrate the role of any given neural structure in any particular function: electrical stimulation provoked the manifestation of the function, whereas a lesion made it disappear. Fritsch and Hitzig placed two fine platinum electrodes on the brain, several millimeters apart from each other, and applied current from a battery (galvanic current). The intensity of the stimulation was weak, just sufficient to produce a little tingling on the tongue of the experimenter. When the electrodes were placed on the anterior part of the brain, the passage of current produced a contraction of a limited group of muscles in the half of the body opposite the stimulated side. Stimulation of a neighboring zone produced a contraction of another muscle group.

Finally, other regions, situated more posterior, were unexcitable. Had the dogma of inexcitability, held until then, arisen because the unexcitable posterior zones of the cortex were more readily accessible to the experimenter? In physiology, method, together with theoretical attitudes, determines results, and it is quite possible that the idea of an omnipresence of psychic functions in the entirety of the brain limited experiments in this domain, and biased their outcome. Fritsch and Hitzig, establishing the first functional map of the cortex based upon objective findings, therefore belonged to the growing generation of researchers, beginning with Broca, who left behind the disciples of Flourens.

The notion of a cortical center for motor activity spread rapidly. Fritsch and Hitzig completed their experiments by ablating defined zones of cortex, previously identified by electrical stimulation. After operation, the animals (dogs) showed signs of paralysis localized to the corresponding member, thus confirming expectations. We will come back to the interpretation that Fritsch and Hitzig gave to this motor trouble. Several years later, the Englishman David Ferrier took on the same problem, with the goal of bringing experimental proof to certain ideas of Hughlings Jackson, whom he admired. Jackson had observed in his patients certain epileptic fits which manifested themselves as convulsions limited to a single extremity and corresponding to circumscribed lesions of the cortex. Ferrier hypothesized that electrical stimulation could be analogous to irritative lesions, those which produced convulsions rather than paralysis. Having chosen to work with monkeys, whose brain is more like a human's, Ferrier used faradic current produced by an induction coil. Faradic current, when applied to the cortex, had the particular ability to produce contractions of long duration, while galvanic current only produced brief contractions at the opening and closing of the circuit. Ferrier could thus observe movements which, he believed, more closely resembled "intentional" movements, the monkey giving the impression of wanting to scratch his belly or grab an object placed before him.[6]

Ferrier's map of the motor cortex of the monkey was too large, extending too far toward the posterior, into the sensory zone. Projected by analogy onto the human brain (for, according to Ferrier, what was true in the monkey was also true in the human), it occupied not only what we now know as the motor area as such but also the entire parietal lobe and part of the temporal lobe.[7] Later, Sherrington and Grünbaum would establish the definitive topography by stimulating the cortex of the chimpanzee, the gorilla, and the orangutan; it was limited to the frontal ascending convolution. In 1880 Ferrier proposed using the results of animal experimentation to intervene in humans, and proceeded with ablation of cerebral tumors. According to him, one should be able to use bony landmarks to get to the cortical zone one wished to lesion, itself determined by clinical symptoms of paralysis or convulsions. For the American C. K. Mills, the functional centers of the motor zone of the cortex represented a mosaic, knowledge of which allowed the neurologist to say to the surgeon: "Cut here or there, or do not touch this or that." From this era one can date the beginning of neurosurgery, marked in particular by the personality of V. Horsley in London from 1890 onward. Neurosurgeons would, moreover, contribute to the problem of localization in the motor cortex by carrying out electrical stimulation of the cortex during the course of surgical procedures. This technique produced the beautiful cortical maps of the human, published from 1937 onward, by W. Penfield.

Electrical stimulation of the cortex validated, by refinement, the results laboriously obtained by the first utilizers of the clinicoanatomic method. The point-by-point topographic representation of the skeletal musculature on the surface of the brain (arising from a speculative conception of motor organization for the most optimistic, from an erroneous conception for others) was entirely confirmed by 1875. Thenceforth, stimulation and ablation were used in a complementary manner in animals: it was thus that Ferrier successively applied the stimulating electrode (which produced a movement) and the cautery (which, destroying a small zone of grey matter, produced a paralysis of the same movement) to the same point on the cortex in the same monkey. What was surprising about this? After all, "if one attributes a motor function to a nerve, because by irritating it one obtains a muscular contraction and by sectioning it one obtains a paralysis of the same muscle, I do not see why motor functions cannot be similarly attributed to a cortical center, given that the phenomena are essentially the same."[8] In their turn,

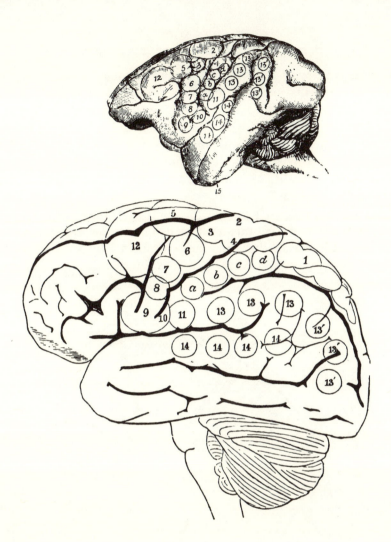

Figure 7 The motor centers according to Ferrier. The regions where electrical excitation produced localized movements are shown on the external surface of a monkey brain (above) and a human brain (below). When region 1 was stimulated, the posterior limb advanced as if to walk. Region 2 corresponded to complex movements of the upper leg, lower leg, and foot. Regions 4, 5, 6, as well as regions *a, b, c, d,* corresponded to movements of the arm and especially the fingers. Regions 7, 8, 9, 10, 11, 14, and 15 corresponded to movements of the face. Regions 12 and 13 corresponded to movements of the head and eyes. It should be noted that the extent of the cortical regions does not correspond to the volume of the muscles in question, but rather to the functional importance of these muscles.

the clinicians, with the support of experimentation and thanks to the improvements in technique of sectioning and histologic examination of the brain, developed a tight correlation between "deficit" motor symptoms (paralysis) and localization of lesions in the frontal ascending convolution. On the other hand, lesions situated outside this zone, even when they were quite extensive, did not produce motor symptoms. P. M. Charcot and Pitres, neurologists at the Salpêtrière, thus established the criterion of "double dissociation," which is still a basic principle of neuropsychology: one cannot confirm the role of a neural structure in a particular function unless the disappearance of this function is produced by a lesion of the

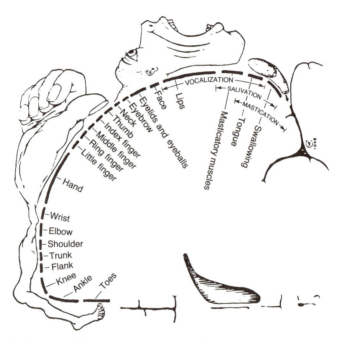

Figure 8 Topographic representation of the movements produced in man by electric stimulation of the motor cortex. The dimension of the body parts of the "homunculus" correspond to the surface of the cortex involved. One should note the relative importance of the cortical surface devoted to the muscles of the hand and face, which Ferrier had already observed in detail. In this drawing, from Penfield and Rasmussen (1957), the "homunculus" is represented on a schematic vertical slice of the brain, at the level of the frontal cortex.

structure in question and not by a lesion of a neighboring structure, and *vice versa*.

The motor zone of the cortex, defined by pathology and delimited by electrical stimulation, was considered as an organ of a special type from 1880 onward. In 1874 the Russian V. Betz described there "giant cells," visible in dogs, monkeys, and man, and which Ferrier quickly considered as being the support for the motor function. Sherrington, using some of Goltz's decorticated dogs, noted that the destruction of this part of the cortex was associated with the degeneration of the pyramidal tract. From this, he came to the conclusion that the motor region of the cortex and the motor region of the spinal cord (the region containing motoneurons) were tightly connected. Thus came about the conceptual union of two zones, at opposite extremes of the nervous system but participating in the same function, the genesis of movement. Reflex movement in the spinal cord and voluntary movement in the cortex? Things were not so simple, and bitter polemics developed regarding the precise role of the cortex. Preferring to avoid all this, Sherrington proposed to call the motor area of the cortex simply the "cord area," a solution which tended toward physiologic pragmatism on the one hand, and concern for associating the role of the cortex with a well-defined motor function, that of the cord, on the other.

But what might the motor zone of the cortex be doing? Lesions at this level did, in fact, produce a paralysis of the corresponding muscles. But this paralysis was transient, since the motor faculties were rapidly reacquired. Ferrier wondered if it were necessary to conclude, as some authors "who could see no further, in physiologic matters, than dogs," had done, that "the surface of the brain has no real association with motility, that the phenomena subsequent to cerebral lesions are no more than transient functional difficulties"?[9] It was Goltz to whom he was referring. In experiments performed on dogs, Goltz noted that when the spinal cord was sectioned at the thoracic level, a whole series of paralytic phenomena were immediately noted: completely abolished were movements of the hind legs, control of bladder and rectum, and erection. Did this mean that the nervous centers for the corresponding muscles were located in the brain? To the contrary, it was known that these functions had their center in the lumbar region of the spinal cord.

If they were in a state of "apparent death" after such a section, it was because they were victims of a remote effect. Goltz wrote that the error of those who had described the effects of cortical lesions (in particular Hitzig) "consisted of the fact that he considered as deficit phenomena that which was in large part merely phenomena of arrest, phenomena characteristic of the first stage of the lesion which quickly improved."[10] That which was true for paralyses produced by sectioning the cord was also true for hemiplegias after a cerebral lesion, according to Goltz. All one needed to do was wait long enough in order to see that the residual deficit produced by the lesion was minimal, or even nonexistent. "After a lesion of the so-called motor zone of the brain . . . one often sees paralyses of the opposite half of the body. These observations are true; they *do not prove*, however, that the surface of this region is the organ which is indispensible for voluntary movements."[11]

The problem, then, was one of the definition of voluntary movement or, even more precisely, a problem of the means to be used in order to determine a specific involvement of voluntary movements. Goltz stated that a cortical lesion did not produce any paralysis of the voluntary muscles. "What the animal has lost with such a lesion," translated Soury, "is the ability to initiate an action, in a smooth and sure fashion, of precisely those groups of muscles whose movement would be necessary to reach a goal, to achieve an end that was desired in a given situation."[12] Was this difficulty the key to the entire explanation? Animals without cortex had "machine-like" gestures. Ferrier, as well as Jackson, thought the same movements (produced by the same muscles) might be the result of several different causes, and be represented in several different centers. "Those who implicate consciousness, and what we call voluntary . . . are the only ones which are of necessity paralyzed by the destruction of the surface, while those which are described as being automatic, instinctual or responsive . . . are organized more or less independently, and completely in the centers subjacent to the surface."[13] Ferrier cites an experiment of Goltz, moreover, as proof: a dog without a cortex used his paw normally for the automatic act of locomotion, but was incapable of using it in a spontaneous gesture such as holding a bone in order to chew it.[14] Thus, according to Ferrier, gestures executed in a certain context were voluntary, while others were not, which explained why only the former were altered by a lesion of the cortex. In the frog more than in the dog, and in the dog more than in the monkey, movements were relatively independent of the cortex. In man, however, it was often found that, after a cortical lesion, automatic movements were conserved while voluntary movements were abolished.

From Ferrier, then, we can date the idea, still well-entrenched in current neurologic thinking, of a duality of motor function, where cortical motor function, conscious and voluntary, stands opposite subcortical motor function, unconscious and automatic. We can certainly trace the origin of his idea back to Willis, but Ferrier was the first to give it an experimental basis, and to discuss it in anatomic terms. There was a small circuit where movements such as gait were organized, in a purely automatic fashion. This circuit did not reach the cortex, but was limited to the striatum and the "subvoluntary" centers. The large circuit was reserved for the execution of voluntary movements, and involved the cortex. A lesion of the large circuit in the dog produced few effects because the small circuit continued to function. In man, by contrast, the small circuit was incapable by itself of serving as a replacement, and the paralysis of voluntary movement was complete. The question of automatic-voluntary dissociation in man, to which we will return later, exceeds the problem of the motor cortex *per se*, if we mean by this the zone delineated by electrical stimulation. The destruction of this zone in man produces a complete hemiplegia, a flaccid paralysis of the extremities of the other side, not only blocking any response to orders of the will but also any participation in automatic actions. When a partial recovery of movement occurs, it permits the execution of voluntary movements and automatic movements as well. The motor zone of the cortex is, therefore, a transmitting organ, which itself receives orders from elsewhere.

Whatever extent was attributed to the motor area of the cortex around 1880, it is interesting to consider the hypotheses invoked around the same time to explain the motor cortex's intervention in the production of voluntary movements. Was the excitable cortex really a "motor center"? Very shortly after its discovery, Hitzig, while observing dogs with lesions of the motor area, was struck not by the paralysis (which was known to be of little importance) but by the maladaptation of the movements to their goal. The paw on the side opposite the lesion slid along the ground during walking, which caused the animal to fall; when the animal stood still, the same paw rested on its dorsal surface rather than on its plantar surface. Hitzig's conclusion: the animal had lost its sense of the position of its extremity, or "the faculty of forming ideas, or representations . . . of this extremity."[15] In short, the difficulty was not motor, properly speaking, but rather sensory in nature. It was certainly not an anesthesia in the usual sense,[16] but the loss of a particular sense, the "muscular sense," of which the motor area represented the final station. In other words, the muscle was still receiving motor impulses, but it was the mind which no longer received information on the state of the muscle.

How could this affect movement? We should not forget the role that Spencer's thought played; his *Principles of Psychology* were reissued in London in 1870. For Spencer, any voluntary act was necessarily preceded by the reliving of a sensation. Sensory impressions felt during the prior execution of an act were relived "as an idea": it was this idea which would then determine a muscular contraction such that the resulting act would exactly reproduce the original impressions. This was the theoretical basis, whose influence on the physiologic constructions of Setchenov and Pavlov we have already studied. In London, a direct disciple of Spencer's, C. Bastian (the executor of Spencer's will, actually), also aware of the work of Fritsch and Hitzig as well as of Ferrier, elaborated an interesting model which gave a physiologic reality to his teacher's thinking. In his first paper (in 1869, a year before the experiments of electrical stimulation of the cortex), he thought it possible to show that "the brain is assisted in the execution of voluntary movements by guiding impressions of some kind," which were different from those produced by cutaneous or deep sensation. These "unconscious" or "unfelt" impressions constituted the "muscular sense" (of which Hitzig would write a year later), and it was these which influenced the initiating center for voluntary movements.[17] After 1870 Bastian no longer had any doubt that the muscular sense ended in the motor area, which, he felt, should no longer be considered "motor." Only the neurons of the spinal cord were "motor," while the cortex was a *kinesthetic* center, and if its destruction produced a paralysis, it was because of the impossibility of evoking the image of movements, and as a result, of commanding them.

Following Bastian's work, other similar theories appeared which imposed the idea of a sensory nature on the motor area. This tendency, which culminated around 1890, can without doubt be interpreted as a return, in force, of empiricism and sensualism: nothing occurred in the brain which did not have a sensation as its origin. According to Hermann Munk (a student of Müller), will and voluntary movement were convenient, perhaps useful, ways of speaking, but were without physiologic reality since disorders of voluntary movement occurring after a cortical lesion were merely a type of anesthesia. For Munk, the cerebral cortex contained only sensations, perceptions, and representations; there were no motor centers, only motor images, of a purely sensory nature. "As soon as the image evoked by association pathways attains a sufficient degree of intensity, movement manifests itself of necessity unless it is otherwise inhibited."[18]

Thus, the domain of physiology saw the development of an explanation which was very close to that which reigned over the psychologic domain at St. Petersburg. Pavlov's "cortical analyzers" recall the "sensitive sphere" which Munk placed in the cortex as the origin of voluntary movements. The sensitive sphere produced movements in the part of the body from which it received sensations, following a scheme similar to a reflex, in the same manner that the visual sphere produced eye movements and the auditory sphere produced movements of the ear. The difference between this and the reflex proper was that the execution of the movement could be varied according to the eliciting sensation. The motor zone of the cortex had no direct motor function, but "consisted of organs at the service of a special memory, the memory of movements, and constitutes a sort of cerebral register which stores the information acquired by the sensory organs."[19] This information lay in the cortical sensory sphere corresponding to the extremity which was the point of departure for its acquisition, and the excitation of this sensory sphere produced movement by the evocation of a "commemorative image." Finally, the destruction of this same sphere produced a paralysis by the suppression of prior impressions. For the Viennese S. Exner, a contemporary of Munk, pathologic case

63

material provided evidence that cortical motor regulation took place under the conscious control of the sensory organs. Thus, a patient with an anesthesia of the arm could not clench his fist or hold an object if he were not controlling his movements by sight; if he were to close his eyes, he would behave like a patient with a paralysis. A blind eye squints, a deaf person is only mute because his cortex does not receive the sounds of language. Motor deficits from cortical lesions were, therefore, only pseudo-paralysis, linked to an involvement of "sensomotility."

If, as we will see, these theories displeased Ferrier, they were nonetheless consonant with those of Sherrington. The latter, in order to test experimentally the role of sensation in the production of movement, had the idea of sectioning in the monkey those posterior (sensory) roots of the spinal cord which correspond to a forelimb. This operation rendered the extremity completely insensitive but, according to Sherrington, it also made it completely immobile, even though the motor roots were still intact. When the monkey walked or climbed, the arm deprived of sensation remained paralyzed; it was also not used when food was presented to the animal and the normal arm restrained. Only a few uncoordinated movements in the affected arm could be noted when the monkey became agitated or attempted to escape. During the four months that one of the animals was observed, however, prehension (that is, movements involving the hand and fingers) was never noted. As Sherrington stated, it was precisely the movements most largely represented in the cortex that were most afflicted, or even totally abolished. "Although we are aware of the danger of introducing terms relating to consciousness into descriptions based almost solely upon motor reactions, we believe that we cannot more lucidly state the condition of these animals than by saying that the volitional power for grasping with the hand, etc., had been absolutely abolished by the local loss of all forms of sensibility."[20] However, Sherrington made the important observation that electrical stimulation of the cortex corresponding to the desensitized arm produced movements similar to those which were usually observed, including movements involving the fingers. Sherrington concluded that a profound difference existed between the voluntary production of a movement and the effect of electrical stimulation of the cortex.

These observations, which confirmed and even surpassed Bastian's hypothesis (in the sense that they showed that not only the cortex but also the sensory pathway which connected the periphery with the cortex was involved during voluntary movement), had undeniable powers of persuasion, given how closely they followed from the theory which inspired them. They were only disproved much later.[21]

There was a motor theory, ardently defended by Ferrier, which limped along parallel to the sensory theory. This motor theory could only find support in the work of the Italian school, which J. Soury rescued from obscurity in his book of 1891. The motor theory, unlike the sensory theory, was lacking in experimental arguments. L. Luciani stated that voluntary movements were the objective expression of modifications in consciousness which, taken as a whole, constituted the will! "Thus, we should not call the cortical centers simply *motor*, but rather *psychomotor*."[22] They were connected with "psychic" centers, and not with sensations. How could a rational mind be satisfied with such an explanation, an astonished Soury wondered. These "so-called psychic centers" resurrected a "metaphysical legacy" stripped of all experimental or clinical foundation. According to the formulation of the Italians, who greatly admired the work of Goltz, it was the cortex as a whole which participated in this psychomotor activity. The reappearance of almost normal motor function, except for "motor ideation," in a dog without cortex was explained by a partial replacement of function by subcortical grey matter nuclei, such as the striatum. In other words, the cortex was only motor because it contained the site of ideas and memory, residues of perception as well as sensation. This cortex, which here became the organ of psychic function, seems close to Flourens's organ of perception and wishing—a conception which would erase practically all the efforts, and all the successes, of the theory of cerebral localization.

In his work on the motor functions of the brain, François Franck proposed that the motor cortex should not be considered as a motor organ or a sensory organ. First, he noted that the paralysis produced by its destruction resolved nearly completely in animals such as the dog, but much less so, if at all, in animals "with the most advanced cerebral organization" such as man. Franck thus joined Ferrier in thinking that the more movements were under the control of the will, the more marked and long-lasting was the paralysis resulting from a lesion of the cortical centers.[23] Several hypotheses might explain this phenomenon of greater or lesser functional replacement: the cortical center of the contralateral hemisphere, cortical areas bordering the lesioned zone, or subcortical centers might take over the afflicted function, with greater or lesser degrees of success. According to Franck, however, it was a "perfecting" of the action of the spinal cord which allowed one to explain the motor rehabilitation observed in certain species. As Pflüger and Vulpian had shown, the cord was itself capable of important work in coordinating and directing movements. Its role as replacement occurred as an inverse function of the importance of the pyramidal tract which transmitted the influence of the cortical motor zone to it. In animals such as the rabbit or the dog, where the pyramidal tract is thin, the spinal cord rapidly took over the movements suppressed by the cortical lesion; in species where the pyramidal tract is an important structure, such as man, this suppression would persist longer, even indefinitely. In this conception, the motor cortex was only an organ of control, holding sway over the cord, which was the true motor center. The cortex was the organ of will, a *point of departure*, and not the movement center. It was not the primary source of motor excitation, but it contained the organs which transmitted these excitations to the true motor centers, which were the brainstem and the spinal cord. "They solicit action, they determine its sense, but they are not the agents of execution."[24]

An intermediary step, the motor cortex was thus not "the explanation" of voluntary movement. Perhaps, after all, its discovery had only pushed the problem of the command of movement higher still, and permitted the supporters of the reflex theory to find a new site. It was in this spirit that Franck avowed, "Better than all others, the reflex theory appears to explain the multiple effects of cortical excitations. Better still, we find that the only neural mechanism known at present, and to which reactions of cortical origin can be attributed, is the reflex. This in only an assimilation, but it excludes the primitive conception of nervous organs, in and of themselves motoric."[25] But was this exclusion in fact necessary?

5

The Motor Idea?

he motor cortex is the exit through which messages destined for subcortical motor structures, in particular the moto-neurons of the brainstem and spinal cord, must pass.[1] It is not surprising, therefore, that a lesion which interrupts the pyramidal tract at its origin produces a complete paralysis of one half of the body. This paralysis, however, does not necessarily imply that other stages of the motor process (those prior to the execution), though now deprived of the means of expression, might not persist upstream from the motor cortex itself. If one were to accept that the cortex's contribution to the motor process was more than just the completion of a circuit which only made excitations arriving from the periphery reverberate, one might well envisage the intervention of cortical areas other than the strictly defined motor cortex in the command of movements.

Caught in a sort of infinite regression, we must question the mode in which these areas of a new, higher-order hierarchy function. Do they still obey some stimulus coming from the exterior? Or are we finally approaching the *primum movens*, that ideal localization where consciousness exists alongside its own incarnation? This is an impossible question, since it allows no outcomes other than mechanism or dualism. The hypothesis of a psychophysical "parallelism" suggesting that consciousness and cerebral function belong to two separate levels without causal relationships between them, is, after all, only the expression of a soft dualism, bringing the two classical theses back to back. As for the notion of "emerging" properties, which would have matter beyond a certain level of complexity acquire new, "nonmaterial" properties, this is, at least for the domain that interests us, a renewed formulation of vitalism. We will give it some consideration in a later chapter.

We must, therefore, accept that neither anatomy nor physiology constitutes an acceptable explanation of behavior. These are, in fact, merely the limitations imposed on any explanation, in the sense that no hypothesis concerning the origin of voluntary movement can proceed without taking them into account. Pavlov dealt with this difficulty by decomposing mental activity into elementary units, each constituting an atom of the psyche and corresponding to a neural structure with its efferent and afferent pathways. In order to explain certain paradoxical responses of his animals when placed in complex situations, Pavlov invoked hypothetical inhibitory mechanisms that permitted the dissociation of response and stimulus. Depending on the situation, a stimulus which was normally excitatory could become inhibitory; behavioral responses thus depended on a complicated algebra, wherein different associated stimuli each brought their weight to bear at a given moment.[2] As Merleau-Ponty noted, "Pavlov's necessity to correct, at each instant, one law by another proves beyond doubt that he has not discovered the central point of view, from which all the facts might be coordinated." His physiology was an "imaginary physiology," functioning, under the cover of objectivity and based on an *a priori* theory regarding the nature of the "physiologic events" which transpire inside the brain.[3]

Merleau-Ponty, placing himself in opposition to Pavlov, embraced other, equally extreme, theses about cerebral function. These theories were, in fact, those of K. Goldstein, who had derived them from observing difficulties experienced subsequent to cerebral lesions in man. For Goldstein, pathologic lesions of the cortex, in addition to producing localized difficulties, simultaneously involve all the domains of function (action, perception, language, etc.). However, "a domain of function is never completely abolished";[4]

rather, in each domain, certain functions remain possible. What characterizes the state of a patient suffering from a cortical lesion is a "fundamental, unique modification" which results in a global alteration of behavior. "One might say that whenever the patient must make an abstraction from concrete facts in order to act, whenever he cannot refer to something except by imagination alone, he will fail."[5] Anything that forces him to go from the sphere of the "real" in order to obtain access to the sphere of "thought" will lead to failure. As regards movement, Goldstein speaks of a failure of voluntary behavior, with conservation of that part of the act which is directly determined by the situation." In sum, "there is the loss of independence, and a stronger adherence to the surrounding milieu."[6] Goldstein describes this fundamental modification of behavior as a "loss of the categorial attitude," an attitude which normally permits one to abstract oneself from a context in order to attain the level of representation and signification.

This conception bore the mark of *Gestalttheorie* then in vogue in Germany. Goldstein, who himself recognized the relationship between his own ideas and this theory, deliberately emphasized certain clinical observations of the patient during examination; he noted, among the symptoms characteristically produced by lesions of the cortex, "a more pronounced deficit in the execution of isolated movements as opposed to the execution of global movements; the inability to distinguish a detail in a picture, while the latter can be recognized as a whole; the inability to pronounce a word in isolation from the context in which it usually appears." "The disintegration," he continued, "occurs in a direction which leads from a more differentiated state . . . toward a more global behavior."[7] The patient had in some sense lost the capacity to extract the "figure" (isolated gesture, detail, or word) from the "background" of his activity when facing a situation. The focal symptoms which resulted from this were only the secondary effects of the main difficulty. It is thus not surprising that cerebral localization occupied little space in Goldstein's theory: symptoms allowed one to localize the lesion to a certain extent, but certainly not to localize function, to the extent that the normal function of the brain could not be deduced from the function of the lesioned brain.[8] Pathologic behavior is a particular and autonomous entity created by a lesion, producing a new form of adaptation to the world which should not be interpreted as the negative image of normal behavior.

Another formulation of this global alteration of behavior after a cortical lesion was given by E. Cassirer. Although he was introduced to neurology by Goldstein, Cassirer approached the subject as a philosopher conscious of the historic sources of his theory and, on the whole, in a less idiosyncratic manner than Goldstein. By taking the Kantian notion of *symbol* as a reference point, Cassirer thought that a large number of deficits in neurology could be explained by an inability to appreciate the symbolic content of objects and situations. Thus, objects could be seen, but not perceived or recognized, or even utilized. In the same fashion, simple or complex gestures might not be executed, not because of some sort of paralysis, but because they could not be "represented." Cassirer said, "There is a form of action which consists of a simple motor activity: a reaction that might be called 'mechanical', immediately unleashed by a given external excitation. There is another form, made possible only by the formation of a determined conception, that of the goal of the action, which must be anticipated and represented by thought. Now, in this latter type of activity, we can always find the trace of a form of thought closely related to verbal thought, and which we may . . . include with it under the concept of *symbolic* thought. Most of our movements and voluntary actions contain a symbolic element of this sort."[9] The pathology of cortical lesions, according to Cassirer, would thus be a pathology of "symbolic consciousness," a conception which, in sum, was rather close to that of Goldstein's loss of the categorial attitude. The specificity of cortical function was affirmed by the type of behavioral disorganization produced by a lesion. The cortex was not a simple supplementary relay which processed external information; this sort of processing could only give rise to an *image* of an object, and not to a *representation* of this object. The image was directly related to the sensory stimuli which constituted the object, while the object and the representation did not have any relationship except on the symbolic level. Between the image and the representation was a gulf which separated consciousness of the object from the object itself.

Cassirer's conception, which gives the cortex a "symbolic function," singularly illuminates certain routine clinical observations, in particular the famous automatic–voluntary dissociation of the neurologists. Jackson, in his papers of the 1870s, provided more than one example: certain patients were incapable of sticking out their tongues when asked to do so, but executed this movement without difficulty in order to lick their lips. Difficulties of this sort were often seen in aphasics. One patient, incapable of speaking or only able to produce a stereotyped phrase when questioned, retained the ability, up to a certain point, to produce words. "We ought rather to say that words come out of him automatically on fit occasions."[10] For example: "Take care!" in front of an infant about to fall, or "That's a lie!" in response to an affirmation. But the patient could not "say what he uttered, and could not repeat out of context the phrase that he had just spoken under the influence of a certain situation: these were words, and not speech. Jackson concluded that "in the speechless man, the most automatic use of words is quite intact."[11] The patient, who is often reduced to swearwords or habitual sayings which come out "automatically," has thus lost the ability to form propositions, to adopt what we might call the "predicative" attitude. "It is not merely a question here of the inability to pronounce propositions, but rather an ability to form them . . . therefore, we do not say that the speechless man has lost his words, but only that he has lost the words which express the possession of language."[12] Everything is a question of the level of mental activity; the lower level of action or language might persist intact, while access to the higher level, which permits the use of gestures or words in acts or speech expressing thought, has become impossible.

Although Jackson acknowledged the predominant role of the left hemisphere in the processes responsible for language (in fact, this had been known since Broca's work around 1865), he refused to localize in any precise manner a "speech zone." "Whilst I believe that the hinder part of the left third frontal convolution is the part most often damaged (in aphasics), I do not localize speech in any such small part of the brain. To locate the damage which destroys speech, and to locate speech, are two different things."[13] At this time, however, localizationist theories held sway: clinicoanatomic observations and the results of electrical stimulation of the cortex provided ample justification. The notion of localized, specialized centers for a given neurologic function converged toward the even more established idea of a circuit, imposed by the studies of the reflex mechanism which were then at their apogee. What Jackson sensed and Goldstein reaffirmed—that a global decomposition of behavior, rather than the loss of localized function, resulted from a cortical lesion—suffered from a lack of proof. On the other hand, a model of functioning based upon centers connected by association pathways not only benefited from the obvious, material analogy of a circuit but was also closely linked to the conceptions of a rising psychology. The psyche was thought to result from associations of images, themselves elaborated from sensations coming from the exterior. The arrival of a given sensation evoked the memory of other sensations, according to the logical laws of similarity, contiguity, complementarity, etc. The association thus created gave rise to an idea which, in turn, was associated with other ideas, and so on. Associationism became an elaborate variation of sensualism, since it allowed one to explain mental activity as beginning with sensation. However, sensation played more than the simple role of external contribution or building material: it was the catalyst which precipitated and maintained the chain reaction of mental events. These conceptions resemble those of Spencer and his heirs, particularly Pavlov, who hoped for an objective approach to cerebral mechanisms using this theory.

Although associationism did not directly favor the concept of localization, neither did it deny it. It led the concept down a rather narrow, set path: in the cortex, one should only be able to find projection centers for sensations, in other words image centers (auditory, visual); on the motor side, one should only find similar image centers, but this time for motor or kinesthetic images. Complex mental functions had no centers of their own, but rather arose from the relationships, via neural association pathways, between different image centers. Anatomic localization in the strict sense (the sense that Gall and Broca had given it) was only the static part of a broader, dynamic whole which made up the functional circuit. According to Carl Wernicke (whose thesis on aphasia appeared in 1874, the same year as Jackson's article), the static part of the circuit played the role of memory, containing the memories of sensory impressions which could resurface by themselves, independent of the stimulation which first produced them. All the rest, in contrast, "all that goes beyond these most elementary functions, such as the flow of different impressions in a concept, thought, or consciousness, is the product of the activity of masses of nervous fibers which connect with each other in different places in the cortex, making up what Meynert called the "association systems.""[14] The association pathways were less resistant to the passage of nervous excitation the more frequently they were used: "It is thus that reflex movements occur more easily when the same pathways are utilized with a greater frequency. It is thus with the exercise of certain movements, such as playing the piano."[15]

Such would be, therefore, the anatomic and physiologic basis for voluntary movement, the simplest of the phenomena of consciousness, according to Wernicke. "Voluntary movement does not take place immediately after stimulation, it occurs thanks to visual memories, the results of prior sensations,"[16] which might be evoked by a new stimulus, or might simply be residual. Spontaneous movement "is distinguished from reflex movement by its enclosed . . . intentional form, in short by a form of movement preformed [by] the ideal representation of the movement to be executed."[17] Thus, one could draw a diagram explaining the appearance of a spontaneous movement: A visual sensation first reaches the occipital cortex, and leaves a visual memory there; at the time of a new sensation, the quasi-latent sensation comes into action, and propagates itself to the motor cortex by an occipito-frontal association pathway; upon arriving there, the stimulus produces the execution of the movement. This group of pathways "is able to explain spontaneous movement by a reflex process."[18] How does the motor representation constitute itself in the frontal cortex? Under the influence of truly reflex movements which, when they were executed, left corresponding motor representations there. In the child, before the development of consciousness, movements, including those involved in speech, are of both an excitatory and reflex nature: they allow the amassing of stocks of motor images which will subsequently be associated with sensory memories in order to bring about spontaneous movements, an achievement that becomes all the easier as the individual gets into the habit of making these associations more frequently.[19]

For Wernicke, the only truly scientific definition of "free will" coincided with the conception of "psychic reflexes." This associationism, at least, merited its name. Wernicke was the most famous of those "diagram-makers" for whom Cassirer had harsh criticism. Wernicke was among those self-convinced empiricists who "believed that they conformed strictly to the facts, and that their observations were imposed on them solely by observation of the immediate," when in fact they had subordinated those conclusions from the start to "preconceived theoretical notions."[20] The double model—localization + association—still functions effectively in clinical neurology, and is what remains of this work. Wernicke's aphasia, produced by a lesion located in the temporal lobe (a "sen-

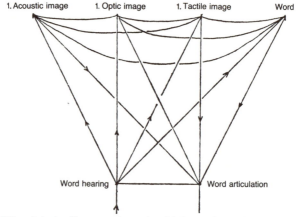

Figure 9 Wernicke's diagram (1903) which explained how certain forms of aphasia appeared after a lesion involving the association fibers of the white matter while the centers remained intact. In this diagram, the image centers (acoustic, optic, tactile), which were stimulated by hearing words, contributed to the formation of the concept and of the word, which would then be articulated. A lesion of one or the other of these pathways would produce transcortical conduction aphasia. One of the possible cases is "word deafness": Language cannot be understood by the verbal channel, while it can be understood by the visual channels.

sory" type of aphasia), and the "motor" aphasia of Broca, are clinical entities that are not questioned.

In the motor domain proper, the systematizing of difficulties took longer to emerge, but used the same approach. Liepmann, who described "apraxia" in 1900, was Wernicke's assistant for four years. It is interesting to note *a posteriori* in the literature of the time a number of observations which constituted potential descriptions of apraxia. The observations of automatic–voluntary dissociation, in particular, or even what certain authors called "motor asymbolia," were not very far off. What was missing in these reports was a unifying concept, a diagram which might, at least as a first step, have explained the nature of the set of symptoms.

The patient who allowed Liepmann to describe apraxia was an imperial councilor, forty-eight years of age, hospitalized on a psychiatry service for "dementia." He presented with a motor aphasia but, most strikingly, manipulated objects in an absurd fashion. A more extensive examination revealed that in fact the difficulty was limited to the right arm, the one the patient used spontaneously. This arm had no suggestion of paralysis: all movements were possible, muscular power was preserved, and activities of daily life, placed in their normal context (using a spoon during a meal, etc.), were accomplished. In contrast, when the patient was asked to perform with his right hand certain meaningless gestures (point to your nose, make a fist), or gestures taken out of their context (show how you use a harmonica, or a brush), the patient failed completely. The gesture was at best suggested, even though the patient made obvious efforts proving that he had understood the instruction perfectly well. In fact, if the right arm were held by the observer, all the movements were rapidly and correctly carried out by the left arm. When a gesture required the coordinated use of both hands, the right hand prohibited the execution of the gesture, even while the left hand was responding correctly. For example, when the patient attempted to pour water into a glass, the left hand took the pitcher in order to pour, while the right hand was bringing the glass to the patient's mouth.

Since neither the comprehension of the instructions, nor the motor execution proper, was defective, one was forced to locate the difficulty which produced apraxia at another level. Conforming to the teachings of Wernicke, Liepmann placed it between the visual sensory and auditory sensory memories, on the one hand, (which, awakened by the examiner's instructions, allowed the patient to construct "the intention") and the motor and kinesthetic memories, on the other (which would have allowed the patient to transform this intention into action). Sensory memories were localized to the cortical visual and auditory areas, and somesthetic memories in the motor cortex and the regions which surrounded it (called the senso-motorium by Wernicke), so apraxia must involve a disconnexion of the association pathways between these various cortical regions. In order to explain the unilateral nature of the difficulties noted in his patient, who was apraxic only in his right hand, Liepmann offered the hypothesis that the disconnexion must have involved not only the association pathways in the interior of the left hemisphere but also the pathways connecting the two hemispheres (the corpus callosum). It was in this fashion that he explained how the visual and auditory memories located in the right hemisphere were unable to replace those of the left hemisphere at the level of the senso-motorium on that side. The death of the imperial councilor allowed the essential verification of this hypothesis, since his brain had a lesion in the subcortical region of the left parietal lobe (explaining the disconnexion between visual memory and the motor centers on this side), as well as a lesion involving the corpus callosum. Thus, a purely psychologic explanation of intrapsychic function (associationism) could be directly transposed into anatomic terms. Liepmann's diagrams, in his later articles, included the names of anatomic structures and the localization of lesions superimposed on his schema of associations between ideas.

Voluntary gesture was thus the endpoint of the intrapsychic and intracerebral trajectories along which information arriving at the image centers via the sensory pathways traveled. The verbal pathway itself belonged to this category, its disconnexion from the motor center explaining the inability to execute gestures upon verbal command. In 1905 Liepmann stated, "I have shown that manual agility arises from the collaboration of numerous cerebral territories with the hand center . . . In particular, it is the left hemisphere hand center associated with the rest of the brain, and in particular with the left hemisphere, which is practically irreplaceable. Thus, lesions located on the left will be particularly noxious for the praxis of all the body parts susceptible to movement."[21] The notion of the functional duality of the hemispheres, already well established for language after Broca and Wernicke, would therefore also apply to the voluntary motor function of the entire body; this observation of Liepmann, crucial for the understanding of certain aspects of motor organization, and in particular for the relationships between movement and language, is still today the object of much work.

The tribulations of the concept of apraxia in the neurologic literature are interesting to contemplate as a way of understanding the evolution of ideas on the production of voluntary movement.

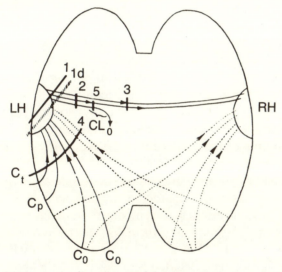

Figure 10 Liepmann's diagram explaining apraxia (1920). The outlines represent the two cerebral hemispheres, connected by the corpus callosum. The motor center for the right hand is located in the left hemisphere (*LH*). Information coming from the sensory association areas of the left hemisphere (*Co, Cp, Ct*) result in motor activity of not only the right hand but also the left hand by the intermediary of the pathways of corpus callosum. The pathways leading to the right hemisphere directly (dotted lines) are considered to be accessory. Lesions placed in various parts of the diagram produce either a paralysis of the right hand with an apraxia of the left hand (1 and 2), an apraxia of the left hand alone (3), or an apraxia of both hands, predominantly noticeable in the right hand (4).

In this evolution, associationism would continue to play a large part, either as a basis of explanation or as a point of departure for the construction of models which would eventually separate off—or even as an adversary against which opposing theories could organize themselves. In his celebrated treatise of 1914, Jules Dejerine (a student of Vulpian) presented Liepmann's ideas in a simplified fashion. Apraxia, which he classified (perhaps revealingly) among the disorders of intelligence, was, he said, "neither a sensory disorder, nor a motor disorder, strictly speaking." For lack of anything better, the definition he gave was entirely negative: "In the series of physiologic phenomena which the execution of an act presupposes, we know: sensory excitation—mental representation of the object—representation of the act to be accomplished—the awakening of motor images corresponding to this act in an ideomotor center—excitation leaving this ideomotor center and, finally, muscle contraction;"[22] neither disorders of the first two steps nor of the latter constitute apraxia, which is therefore a disorder of the intermediate steps. In the course of those steps, the mind first synthetically conceives the act to be accomplished, and represents it as if it were accomplished; this is the psychologic plan, or "ideational project." These mental representations are then transmitted to the ideomotor center, where they awaken corresponding motor images; finally, the act itself is produced.

The interest—or the danger—in such a scheme is that it allows one to envisage as many degrees of involvement of psychomotor function as one postulates intermediary steps. It was thus that Liepmann distinguished three forms of apraxia which Dejerine illustrated by example: "Let us take an individual to whom one has given the order, or who decided himself, to raise a glass and drink. He knows that he is in the presence of a receptacle containing a liquid destined to satisfy his thirst. It is possible that he does not know that to drink, he must take the glass, bring it to his lips, and incline it progressively. One would say he has an *ideational apraxia*. It is also possible that, knowing what movements he must accomplish, and having in some sort organized his plan of action, he no longer knows what he must do to grab the glass, lift it, etc.; one would say this patient has an *ideo-motor apraxia*. Finally, while knowing that in order to take the glass, it was necessary to contract the muscles which flexed the digits . . . it sometimes happens that, in the realization of the act, the execution does not correspond to the conception. One would then find oneself in the presence of a patient with *motor apraxia*."[23] Thus, concluded Dejerine, there is an apraxia of conception (ideational), apraxia of transmission (ideomotor), and an apraxia of execution (motor).

The associationist or "connexionist" conception of psychic activity implied, as a corollary, a pathology of disconnexion. Was apraxia, often considered a "psychic paralysis," really the result of an intrapsychic disconnexion which isolated the apparatus for the execution of movements from other stages preliminary to this execution? The modern theory of disconnexion, although it poses the problem differently (conferring upon it an even greater descriptive, if not explicative, power), preserves this viewpoint. To consider the patient with a cerebral lesion as an indivisible whole is a source of confusion, according to N. Geschwind: the patient is, in reality, made of assembled parts. According to Geschwind, therefore, posing a question like: "Why is a patient, who understands instructions and whose arm is not paralyzed, unable to correctly execute corresponding gestures?" makes no sense, since it is not the same part of the patient which receives the order and which executes it. We have no difficulty accepting this concept of disconnexion when we are speaking of the inferior segments of the nervous system. When we talk of a patient whose spinal cord was accidentally sectioned and say, "The patient urinated," it is not necessary to say, "The patient's spinal cord urinated." However, we are convinced that this act of urination is involuntary, and only depends upon a part of the nervous system disconnected from the whole. It is not a question, states Geschwind, of replacing a "holistic" approach with an "atomistic" approach toward behavior, but it is sometimes useful to consider an animal or man not as units," but as the union of loosely joined parts."[24] This attitude, according to Geschwind, allows us not only to describe a pathologic state but also to understand why, in a normal man, the different parts of the brain interact with such difficulty. This position, extreme as it is, draws some of its justification, at least at the formal level, from the work of Sperry and his collaborators.[25]

Is it in fact appropriate to maintain under a common rubric

disorders as different as ideational apraxia and ideomotor apraxia? Should we not worry that the desire to unify disorders of motor function will lead to a convenient but simplistic explanation of its normal function? Or that the pathology of disconnexion is merely the logical, but inevitable, conclusion of the associationist schema. These are old questions. Pierre Marie, a student of Charcot's (yet another "diagram-maker"), thought that ideational apraxia could be the result of a global disorder of the intellectual abilities, and not a localized lesion interrupting one or another circuit. This viewpoint, which is more or less written across the antilocalizationist tradition from Flourens to Lashley, ended in separating the two principal forms of apraxia and making ideomotor apraxia alone a disorder of gesture. One could easily conclude that the gestures of a patient presenting with an ideational apraxia are inappropriate, even absurd, with respect to the given task or object toward which they are directed, but are nevertheless executed correctly when they are considered in and of themselves. The deficit in this form of apraxia rests more upon the capacity to use objects, as if their "concrete" identification (that which allows us to use them) had become impossible, while at the same time their "abstract" identification (that which allows us to name them and describe them) is preserved. It is thus more a difficulty in recognizing the pragmatic significance of objects than a disorder of gesture *per se*. This concept remains very theoretical: how could one understand automatic–voluntary dissociation which implies that, in certain situations at least, objects can be correctly utilized?

The other forms of apraxia, on the contrary, better demonstrate an intrinsic disorder of motor function, somehow ending up with its disintegration (in the Jacksonian sense). One could easily imagine a physiology of the beginning of movements, which would be different from that of execution itself but would still belong in the motor sphere. The problem then, would be to find for the representation or "plan" of the action to be executed an approach which owes nothing to associationist theory and which would, at the same time, liberate apraxia from its domain. Clinical manifestations common to the motor and ideomotor forms, noted very early but then relatively neglected, allow us to sketch this approach. In 1908 Wilson had noted in several patients presenting with apraxic disorders a perseveration of posture, above all during prehension movements. Thus, one of his patients was unable to release an object held in his apraxic hand, and was obliged to remove it with his other hand. Similar phenomena were often described: a patient who had been asked to pour himself a drink from a water jug brought the jug directly to his lips, and kept it there for a long time; another, given the command to light a candle, lit the match and held it until the flame burned his fingers. This perseveration could extend to the total loss of motor initiative, the patient resting stock-still and immobile.

For the American neurologist Derek Denny-Brown (a student of Sherrington's), the explanation of apraxia had to be sought outside all disorders of ideation or the will. The lesion which produced apraxia had the effect of destroying an equilibrium in the interior of the motor system between two opposing tendencies which constituted the modalities of "tropism for the environment." Perseveration (which Denny-Brown called "magnetic apraxia") was thus an exaggeration of this tropism, resulting in an inability to remove oneself from a situation, a tendency to "stick" to objects. This hypothesis has the interest, at least, of underlining the relationships between the subject's action and the environment where this action occurs. Multiple sensory information arrives from the environment: a poor synthesis of different modalities, caused by a lesion, perturbs motor behavior in the sense of an exaggerated adherence or withdrawal. Denny-Brown's theory no longer makes reference to intrapsychic processes: it is entirely sensorimotor and no longer ideomotor.

Progress in the anatomy and physiology of the cerebral cortex during the 1950s allowed the distinction to be made between cortical "projection" areas (or primary areas), directly connected to the sensory organs and receiving information from a single sensory modality (visual area, auditory area, etc.), and "association" areas in the cortex, placed downstream from the primary areas. The particular function of the associative areas (which, in fact, occupy the greatest part of the cortex) seems to be receiving information from several primary areas and thus from several sensory modalities at once; in other words, they are "convergence" areas. This concept of convergence is different from the concept of connexion, even though both rest on the same principle, that of propagation of neural information from center to center. The concept of connexion implies that a product of the nervous system (a movement, for example) is constructed from elements which are progressively associated with each other—from the simple toward the complex, or from the concrete toward the abstract—according to the classical diagrams. The concept of convergence, on the other hand, implies that multiple bits of information concerning the same object or event arrive simultaneously at the same neural elements.

From this encounter, still hypothetical,[26] between several sensory modalities, information of order greater than the simple sum of the constituent parts is produced. Let us take the example of an object placed at some distance from the body, and toward which a prehension movement is oriented. The visual information concerning the nature of the object (its form, size, color) is dealt with by the primary visual areas of the cortex; dealing with information on its distance from the body is even more complex, since the apparent size of the object, parallax, the degree of convergence of the eyes must all be taken into account; finally, the execution of the movement must take into account not only the already-mentioned information, and different interpretations of it, but equally the position (with respect to the body's axis) of the gaze which fixates the object to be grasped, the initial position of the arm, etc. Only a very large convergence of information documenting all these indices could explain the constitution of the single reference point toward which our actions are directed in space.

A disorder of one of these associative zones of the cortex can thus produce a profound disorganization of movement, not from the destruction of a communication pathway but from the disappearance of an autonomous function, in itself subsidiary but nonetheless essential for motor behavior. In the specialized literature, this same idea can be found on several occasions: apraxia can be linked to a disorder of the "spatial" representation of gesture (what certain authors consider as "spatial dyskinesia"). The fact that the lesions responsible for apraxia are often located in the parietal cortex, precisely in the region where associative areas implicated in the construction of spatial reference have been identified, gives these hypotheses a certain validity.

The theory of an intrinsic disorganization of motor function as the origin of apraxia is opposed by the theory of a larger disorganization involving all the forms of symbolic activity, according to Goldstein's and Cassirer's formulations. Both, however, reject the associationist model of "the motor idea," and its corollary, disconnexion, which would isolate the center of ideas from the motor center. The theory to replace this, a true approach to voluntary movement, has yet to be found.

6

The Physiology of the Will

*I*f physiologists had difficulty obtaining access to the will and if, in the end, a number of them reacted by refusing to include it in their conception of the neural organization of movement, it was because the will was in the domain of the subjective. Introspection was not among the techniques of physiology: it is, however, only through introspection, using language, that a subject can reveal his or her *internal state*. Should one neglect this, in order to limit oneself to that which is observable and recordable? In that case, movement would become the privileged, even unique, criterion of mental activity, characterized only by its external manifestations. Such was the behaviorist attitude, which reflected the predominance of the reflex model in the explanation of behavior.

Does a physiology of the subjective exist? For to leave the subjective to the philosophers (who, it should be noted, could quite easily do without physiology) would have run the risk of creating several degrees of causality in order to explain cerebral function: psychic causality for internal states (a vague causality wherein physiologic or anatomic bases are not sought) and a firmer, physiologically-based causality for external states. Unless one wished to return to a complete psychophysical dualism, it was best to

84

consider external states as the expression of internal states, which could, moreover, exist independently of this expression. These two levels, internal and external, would then become expressions of the same causality. But to what point should this be traced back? To the point of excitation, following the steps of sensorimotor trans- formation backwards, in the case of a movement precipitated by the exterior? Or as far as the psychic process, in the case of a spontaneous movement?

The philosophers of the will, moreover, had not been indif- ferent to physiology. In the first years of the nineteenth century, influenced by the postrevolutionary "idealist" philosophers such as P. J. G. Cabanis and A. Destutt de Tracy (but rather critical of them, nonetheless), F. Maine de Biran neatly posed the question of which physiologic criteria allowed one to distinguish voluntary from involuntary movement. "If this experimental science [physi- ology] could determine the true organic condition which corre- sponded to a *willed* movement, and clearly differentiate it from that which was not, would one not have found the sign or the symbol of the *will?*"[1] Maine de Biran adopted the physiologic explanations of Bichat, which distinguished organic contractility, without direct relationship to the brain, from animal contractility, which emanated from a motor center. By taking up this idea, he considered the motor center as being able to receive a "provocative impulse" which determined a "forced reaction," but "motor action can also begin directly in this center, or only be determined by impressions re- ceived or originating in its own right."[2] It was thus a question of knowing whether the observations of physiology could explain this spontaneous determination of action. For Maine de Biran only vol- untary contractility should be accompanied by a subjective expe- rience, by a feeling of "individual power of effort or action." The "feeling of effort" is at the center of Biran's philosophy: it was not only the subjective criteria of voluntary action, but also the feeling around which the self structured itself: "If the individual did not *will*, or were not determined to begin to move himself, he would know nothing . . . he would not suspect any sort of existence, he would not even have the *idea* of his own self."[3] I act, therefore I am.

Maine de Biran, however, ended up doubting whether physiology could entirely explain the feeling of effort. The execution of a movement was certainly animal in origin, but the feeling of effort did not originate as an animal force, nor even a simple organic force: it was "hyperorganic" in origin. It is obviously not up to us to decide what Maine de Biran put under this heading, nor what attributes he would have given it in a physiologic context different from those of Bichat and Cabanis. Let us simply hold on to the notion that a feeling of effort was a criterion for the voluntary act: there is, in this, more than a mere premonition, since this criterion, rediscovered or taken up again by psychologists and physiologists, gradually took on a functional significance, and even became operative in neuropsychology. Extending the analysis of voluntary phenomena, Maine de Biran excluded any idea of temporal succession between the willing of an act and the act which resulted. In other words, the intention was no different from the action, and the external state was only the visible part of a unique process. The feeling of effort abolished the dualism of the mind and body: "The self could not exist for itself, without having the feeling . . . of the consciousness of the body." "My body and I are one."[4]

The conception of A. Schopenhauer, the other nineteenth-century philosopher of the will, was much more classical from a physiologic standpoint. It was true that Schopenhauer, while referring to the same sources as Maine de Biran, resorted to the objective method, claiming to practice a "philosophical physiology." Philosophy lent weight to the research undertaken by the physiologists which, left to itself, could only lead to "a materialism as gross as it was stupid," to a "philosophy of barbers' apprentices."[5] According to Schopenhauer, the term "will" took on a very general sense: it was not synonymous with the mind, but almost. It was the "root of the mind" which held an ontologic value, like "effort" for Maine de Biran. Will was a *force of nature*, which belonged to all living beings; to know and understand this force was to know and understand life. For this reason, "we must seek to understand nature through ourselves, and not ourselves through nature."[6] The influence of Bichat is quite obvious: Schopenhauer's will has the same role as "organic life," that of a vital principle around which all the manifestations of life were structured. In this respect, the voluntary act and the involuntary act were both products of the will. Free will was not exercised in this or that action, "but in our existence, in our very being."[7] The criticism from psychologists that such positions encountered can be readily understood. As Buchanan basically stated, "metaphysicists" were considering volition as the result of the exercise of an innate power, the will, about which it was not even necessary to pose questions. The metaphysicist got off too easily, since he contented himself with affirming that a mental operation was only the result of an elementary faculty of the mind!

Schopenhauer distinguished between the will in the general sense (*Wille*), and the will at work in each particular movement (*Wilkur*). The latter characterized the voluntary act, whose execution did not depend directly upon excitation. The voluntary act was the only act to be determined by a *motive*, an excitation, which passed through the brain where it gave rise to an image, which, in its turn, produced the reaction. Motives were therefore mental "representations" and "were located in the brain."[8] It was from this, undoubtedly, that we can date the viewpoint of psychologists like R. Wundt, and above all that of the neurologists of which we have spoken in the preceding chapter.

The mental representation of the act to be accomplished thus became an indispensable step in the series of events leading to a voluntary act. The notion of the *program* of a movement, more restrictive in the sense that it does not necessarily imply the existence of a mental phenomenon, does suggest a series of distinct processes preceding the execution proper. Programming and execution were separated not only by the *time* which determined their succession, but also perhaps by the *distance* between different cerebral localizations.[9]

Motive, representation of the act, motor program: we are, in any event, far from Bastian's kinesthetic image or Munk's commemorative image, which were only, in the end, "delayed" effects of sensations. Representation, on the other hand, was a mental construct, certainly utilizing in part the material of sensation, but possessing its own autonomy. At the extreme, representation owed nothing to the external; it was spontaneous. In 1859 A. Bain said the brain was capable of spontaneous activity. This was not a simple thing to acknowledge: the spontaneous act which was released by the will alone, the representation which was spontaneously created, were examples of "open" systems. In other words, they were not bound by the physical principle of conservation of energy, which required that the effect (movement) was equal to its cause in terms of energy.[10] Reflex movement was obviously the typical example of an energetically closed system. In order to get around this difficulty, Bain acknowledged the existence of a parallelism between the "physical side" of the organization of movement, which obeyed the law of the conservation of energy, and the "mental side," which did not. Although these two sides were inseparably linked, their substances were different, a bit like the three persons of the Holy Trinity, to use Bain's terms. Psychophysical parallelism saved appearances, since it was no longer a question of dualism in the strict sense. But was it really compatible with Bain's professed demand for a "physiologic psychology"?[11]

No matter what status one attributed to the steps preceding the execution of a spontaneous act or, in a more general fashion, to mental representation and the entirety of psychic phenomenon, couldn't an attempt be made to measure them? "Our sensations, our representations, our feelings are *intensive* dimensions, which follow one another immediately over time," as Wundt, the virtual founder of physiologic psychology, put it twenty years after Bain. "Our interior life thus has at least two dimensions; this implies the possibility of representing it in a mathematical form."[12] Two dimensions: intensity and duration; no longer only one dimension, time. We should note that the ideal quantification of Wundt rested upon an essentially materialist base. Since it was a question of measurement, we must first look to determine what, in the internal representation which preceded movement, might be measurable.

Could the feeling of effort, a unique subjective experience which accompanied voluntary movement, be objectified to the point of becoming a quantifiable dimension? This feeling belonged to the *active* production of a movement: it somehow indicated to the subject that he himself was the agent or cause of his own act. It was the "intimate feeling of the cause or force which produced the movement that is the *self* identified with its effort."[13] Maine de Biran clearly distinguished between the feeling of effort and other sensations which might also result in the execution of the same movement. We "feel" the passive displacement of one of our extremities when it is produced by a force exterior to us, but this "muscular sensation" is the result of the movement, and not its cause: "muscular sensation unaccompanied by effort or not caused by the will . . . can only be a completely passive impression, just as the heartbeat is, or just as convulsive movements are. We sense them without producing them."[14] The two—feeling of effort and muscular sensation—were normally linked in an indissoluble fashion, since one was cause and the other its effect.

Maine de Biran also had the intuition that the feeling of effort was, above all else, a feeling of *resistance* to action on the part of the will: "Without something which resists, there is no effort." One might well ask, what would happen if the resistance was such that a movement could not be produced? The feeling of effort would then be dissociated from its normal effect, and would be perceived in isolation. Such a situation might occur in the case of a paralysis, which blocks the effort needed to contract certain muscles. The neurologic literature furnished several interesting self-descriptions of this. The physicist Ernst Mach, who was struck with a right hemiplegia in the course of a trainride, was one of these. At first, when the paralysis of arm and leg were total, Mach noted that the effort of will which he made to move the arm, although clearly perceptible, was not accompanied by any real sensation of effort. Later, after recuperation had begun and the paralysis was less dense, each effort of the will was accompanied by the sensation of having the right hand and foot held down by enormous weights: the sensation of effort was perceived as a sensation of weight or resistance which opposed movement. Another victim of a hemiplegia, undoubtedly less dense than Mach's, the anatomist Alf Brodal, was struck by the considerable force which he had to use when he wished to move one of his paralyzed extremities: "Subjectively, [this force] is experienced as a kind of mental force, a power of will . . . it is as if the muscle was unwilling to contract, and as if there was a resistance which could be overcome by very strong voluntary innervation . . . this force of innervation is obviously some kind of *mental* energy which cannot be quantified or defined more closely."[15]

Analogous observations were reported by the rare volunteers

who experimented on themselves with a total paralysis due to curarization.[16] As soon as he tried to move, the curarized subject felt the impression of an intense effort being brought against an invincible obstacle, as if the body had been transformed into a block of stone or cement. In this specific case, the sensation of effort could only be the expression of the emission of a mental force, linked to the command of movement since, as is well known, curare blocks neural transmission at the neuromuscular junction, and no contraction can come through.[17]

Spontaneous activity proceeding from the mere intention of action, the exercise of voluntary motor activity, thus produced a feeling, a sensation. As with any sensation, it belonged to the subject who experienced it. Thus, in 1878 George Lewes stated: "The psychologist will likewise accept a class of motor feelings such as cannot be included under any one of the five senses—Feelings of Effort, Resistance, Movement, Fatigue which have characters not less distinct than those of Colour, Temperature, Sound, or Taste."[18] Without a doubt, they were no less quantifiable. The problem which then arose, and which gave birth to one of those controversies so common at the end of the last century, was to determine the origin of the sensations enumerated by Lewes, to which we can now add the "sensation of paralysis." The two opposing views appeared irreconcilable and mutually exclusive, even if, later on, they were reassembled into the heart of a "hybrid" theory. At stake in the discussion was once again whether or not the brain was capable of spontaneous activity. Could one experience a sensation in any way other than in response to an excitation coming from the exterior, or from the periphery of the body? One can easily recognize here those questions we have not stopped asking ourselves regarding the source of motor activity, either reflex or spontaneous.

In Germany, the circle of Müller's students deliberately opted for the notion of "sensation of innervation." According to this idea, the subject could sense the motor impulses that the central nervous system emitted to the muscles. These sensations were central in origin, and no longer peripheral, traveling along motor nerves rather than sensory nerves. The two types of sensations were, in fact, associated in the feeling which accompanied a voluntary movement:

the sensation of innervation, which represented the intensity of the effort of the will, was normally associated with another sensation of peripheral origin, which resulted from muscular tension. As Wundt said in 1874, the sensations accompanying the act of willing something were usually fused with "the real muscular sensation, associated by the stimulus which is represented by contraction. In other words, their effects cannot be detected separately . . . except when a total or partial muscular paralysis has deranged or entirely destroyed the other muscular sensations which come from the periphery."[19] This was what Maine de Biran had already basically said, and it is essentially what paralyzed subjects (either experimentally paralyzed, or paralyzed as the result of a pathologic condition) said they felt.[20] The important thing, however, was that this sensation of innervation was going to become the basis for a new theory of movement and behavior and, like any new theory, would call the established theory into question.

Let us call the classical theory the *theory of afference;* it placed a sensation resulting from a modification of the external world at the origin of all movement, and from it were derived all the various reflex theories. Let us call the new theory the *theory of efference.* The feeling of effort does not in any way proceed from an afferent influx into the central nervous system via sensory nerves: it is, to the contrary, an efferent influx carried by motor nerves that we feel. From the outset, however, difficulties emerged. According to Lewes, the theory of efference postulated that the notion of force is a direct revelation of the consciousness, a pure cause (he called this its metaphysical counterpart). "Independent of, and antecedent to all experience, we know the degree of Force, or Effort, which a movement demands: this is a birthright of the soul."[21] Lewes, in fact, had difficulty accepting the ultimate consequences of the theory of efference, to the extent that it implied an innate capacity to estimate the force necessary for any movement. One might well accept that the feeling of effort was a "motor intuition framed out of already registered experiences," as we can have the idea of a circle before having drawn it. However, it would be going too far and too fast to suppose that the notion of force "had been in existence before the movements which it ideally reproduces had originally produced it."[22] This radical critique of the theory of efference was a backdrop to the arguments of its principal detractors even if, as we will see, they engaged in combat on a much more factual terrain.

Among the arguments used by the German school to demonstrate the reality of the sensation of innervation, we must remember those advanced by Helmholtz above all. As he stated, how is it that we perceive the world around us as stable when we move our eyes, given that each movement of the eye sweeps the objects surrounding us across the retina? In other words, how do we distinguish a real displacement of the world around us from a displacement of our retina? The question is not a simple one: if we close our left eye, and someone else passively displaces our right eye by placing slight pressure on the outer edge of the eyeball, this time we see the surrounding world displaced.[23] The difference between the voluntary movement of the eye, where the world seems stable, and the passive displacement of the eyeball, where the world seems unstable, is that only in the first case is the movement due to intention, an effort. What would happen in the third possible case, when voluntary effort was not associated with movement? This is what happens when the ocular muscles are paralyzed: in this case, said Helmholtz, when the eye does not obey the will and remains immobile despite efforts to move it, objects in the surrounding world seem to be displaced in the direction of movement the eye would normally have effected. As a result, "these phenomena prove conclusively that our judgements as to the direction of the visual axis are simply the result of the effort of will involved in trying to alter the adjustment of the eyes." We know therefore what impulses of the will have to be employed, and how strong they must be, to bring the eye into some definitely intended position."[24]

Helmholtz took his observations on ocular paralyses from a work by the ophthalmologist A. von Graefe, who had also noted another interesting finding in this type of patient: When he asked them to point quickly with one arm in the direction of an object placed some distance away on the side of the paralyzed eye, von Graefe noted that the patients pointed too far, and passed the object's position. His interpretation was that the intense effort required to displace the paralyzed eye was perceived as such by the subjects, and ended in an overestimation of the movement the eye needed to effect: pointing their arm in the supposed direction of the eye, the subjects thus produced an exaggerated movement of the arm. The illusion of displacement of external objects, and the exaggerated amplitude of pointing movements, were corollaries of the effort made to overcome the paralysis of eye movements. Better than the simple feeling of this effort, these phenomena objectified the fact that the brain set aside part of its own activity for the purpose of regulating the perceptual consequences of action, or for controlling the precision of movements. The cause of a movement could thus become an object of the consciousness or, much more simply, constitute information regarding its own unfolding.

These two alternatives were absolutely rejected by the partisans of the theory of afference. A bitter controversy ensued, especially in England, after Bain introduced the ideas of Wundt, and then adopted them (at least for a time). He opposed Bastian and his supporters, as well as Ferrier, for whom all sensations accompanying a movement led back to a muscle sense of strictly peripheral origin, without any central component. In 1900 Sherrington provided them with crucial arguments by demonstrating the role of "sensory" receptors in muscles, tendons, and joints in the genesis of sensations associated with movement, particularly in the perception of the respective positions of an extremity and the position of extremities with respect to the body. Thanks to "proprioception" (a term coined by Sherrington for this new sense), sensations of movement were reintegrated, at least provisionally, into the centripetal orthodoxy.

Arguments for the theory of afference were taken up at length by the American psychologist William James in the chapter "Will" in his book *The Principles of Psychology*. His first argument was that it was not necessary to postulate the existence of a sensation of innervation. The *idea* of a movement is the last event which precedes it: why imagine otherwise, since the tendency of consciousness is to minimize complication? To know the effort produced in order to effect a movement would be superfluous and cumbersome. The second argument was that there was, in fact, no demonstration of the sensation of innervation. The intense feeling of effort that the paralyzed subject felt could be explained in a very simple manner:

in forcing himself to contract a paralyzed muscle, the patient in fact contracted other, healthy muscles. It was this contraction which, via the centripetal pathway (the "normal" pathway for sensation) gave him the notion of his effort. In the ocular paralysis experiment, had the other eye, which was covered, been observed during the experiment? It moved, and it was its real movement which produced the impressions felt by the subject. James concluded, "Beautiful and clear as this reasoning (on the part of the supporters of the theory of efference) seems to be, it is based on an incomplete inventory of the afferent data."[25]

Having fallen into obscurity for more than a half-century, the theory of efference suddenly reemerged in 1950. Simultaneously, R. Sperry in Chicago and E. von Holst in Germany published observations on fish and insects which, devoid of any "psychologic"

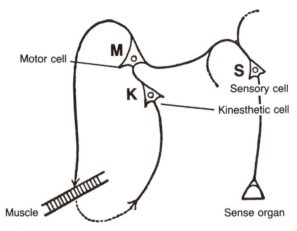

Figure 11 Idealized diagram of reflex organization of movement according to W. James. A sensory cell, connected to its peripheral receptor organ, produces activity in a motor cell, which results in a muscle contraction. This contraction excites a kinesthetic cell, which acts upon the motor cell by inhibiting it. If this did not occur, James noted, contraction would not stop and the muscles would remain in a state of "catalepsy." This diagram includes the notion of "negative retroaction."

It should be noted that the cells and their processes constitute a continuous network. At this time (1890), the fact that communication between neurons was effected by synaptic contacts and not by continuous propagation was not yet known.

94

context, gave the theory its true physiologic basis. In such animals, it is possible to invert vision by turning the eye 180° around the optic nerve, then fixing it in this new position (the other eye is removed). The result of this inversion, readily verified, is that the animal turns on itself indefinitely. How does one explain this behavior of forced rotation? Owing to the turned position of the eye, any movement of the animal entails an excitation inverse of that which is normally expected by the central nervous system (which has kept its connections with the retina intact). When the animal is displaced forward, instead of the normal passage of objects in the visual field from the nasal pole toward the temporal pole of the eye, the animal sees a passage from the temporal pole toward the nasal pole, as if it were swimming or flying backwards. This visual information contradicts the orders sent out by the animal's motor system. For Sperry, any movement (of the eye, the head, or the whole animal) which produced a displacement of the visual image on the retina had to be accompanied by a *corollary* discharge toward the visual centers to compensate for this displacement on the retina. "This implies an anticipatory adjustment in the visual centers specific for each movement with regard to its direction and speed."[26] This is how the stability of the visual field is maintained when an animal moves. After the animal's eye is turned 180°, "any such anticipatory adjustment would be in diametric dysharmony with the retinal input and would therefore cause accentuation rather than cancellation of the illusionary outside displacement,"[27] hence the forced rotation of the animal. Sperry thought that this explanation was the neural basis for Helmholtz's sensation of the effort of the will.

In the same year as Sperry's work, the physiologist E. von Holst, who was associated with the cyberneticist H. Mittelstaedt, came to a similar conclusion by a different path. They posed the question in the following manner: From the swarm of afferent information which bombards a mobile organism, how can the organism distinguish real displacements of objects in the outside world, as registered on the retina (ex-afferences), from displacements that result from the organism's own movement (re-afferences)? In the two cases, movement detectors in the retina are stimulated in the same fashion. Von Holst noted that, without a mechanism allowing for this distinction to be made, no behavior would be possible. As an animal would begin to pursue prey moving toward the left, his action would produce a displacement of all objects on his retina toward the right. This new stimulus would produce a movement of the animal toward the right, interfering with his pursuit behavior! If the animal could, in fact, distinguish displacements of the visual world which are external from those which it engenders by its own action (ex-afference from re-afference), it is because the visual centers which analyze these displacements "know" how to recognize those which are re-afferences, and do not transmit them to the motor centers. How do visual centers have this prior knowledge? For von Holst, every time the motor centers send the command for a movement (an efference) toward the muscles, they also send a "copy" of this command to the sensory centers involved with movement, in this case the visual centers. The model thus takes on an algebraic form, wherein all of the information arriving at the visual centers is given a sign: that which comes from the retina receives a + sign, while the copy of the efferent message receives a − sign. If the information arriving from the retina is an ex-afference, its action, thanks to the + sign, will be directly used by the motor center, resulting in an appropriate pursuit behavior. If, on the other hand, this information is a re-afference, its action will be cancelled by the opposite sign of the copy, in the same way that the superposition of the positive and the negative of a photograph cancel any image: thus, the visual center perceives no displacement and will result in no action by the motor center. As we have seen, Sperry's intuition was similar, but his model was more difficult to generalize. According to von Holst, the reduction of the operational elements to algebraic signs allows one to apply the principle to a great number of situations, from the behavior of forced rotation in a fly whose head has been turned 180°, to the perceptual stability in man during eye movements (Helmholtz's example).

Corollary discharge or efference copy represent two nearly identical formulations of the fact that the nervous system can inform itself about its own activity, and about its own motoric production. It is a question of information which owes nothing to the external environment: it is transmitted from motor centers to sensory centers, that is, in the opposite direction from the sensorimotor pathway, according to the laws of reflexes. Therefore, associated with movement is an endogenous, spontaneous activity of the central nervous system that is capable of producing autonomous organizational forces without recourse to external energy. The central nervous system, the brain, dominates, since it "knows" what will happen as a result of its own action. It anticipates the effects of its own action, in order to maintain an invariant environment. It is no longer the external world which organizes behavior, as reflex theory

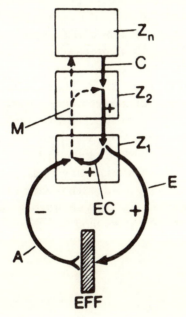

Figure 12 Schematic diagram of the principle of *Efferenzkopie* according to Von Holst (1950). The command center Z_n sends motor orders to the excitation center Z_1, whose efferent output activates the peripheral effector (*EFF*). Simultaneously, center Z_1 generates a copy of the efference (*EC*), of the same sign ($+$) as the efference itself.

Activity of the peripheral effector produces a reafference (*A*), which becomes combined with the copy of the efference at center Z_1. Since the reafference, whose value is ($-$), is the opposite sign of the copy, their effects cancel each other. Nothing is transmitted to the subjacent centers. In the instance where the copy of the efference and the reafference do not cancel each other, a signal *M* is transmitted to Z_n, which produces a reactivation of the motor circuit.

would have it, but rather it is neural activity, at the extreme in a preorganized form, which organizes behavior. The theory of efference produced this sort of reversal in thinking, and von Holst was aware of this when he noted that if reflex theory had been defended for so long, it was because one does not "readily give up a simple theory, particularly when it is regarded as a 'fact' because of its long history, before a better one is available."[28]

Is this new theory better? It is complementary, just as a theory of behavior that is nativist in inspiration can be complementary to one that is empiricist in inspiration. Above all, it allows one to explain how complex modes of function at a high level in the hierarchy can, according to von Holst, "send roots down to the simplest basal functions of the nervous system." The theory of efference abolished the barrier which had separated explanations of behavior on the psychic level from explanations by neurophysiologic mechanisms. The problem of the feeling of effort, and the sensation of innervation, which had so shocked the supporters of "orthodox," that is, centripetal, sensation, was resolved. The principle of re-afference, or the notion of corollary discharge, made the phenomenon commonplace, by designating it as a mechanism to signal spontaneous movement, a subsidiary which contributes to the objectivity of perception by preserving its invariance. The "birthright of the soul" was more a physiology of central command for spontaneous acts.

7

The Physiology of Spontaneity

he notion of a cerebral determination of spontaneous movements and the mental or psychic states associated with them
sooner or later raises several important questions for the
physiologist: Upon what mode of functioning are movements which
are not simple responses to the environment based? Could one
conceive of, and then demonstrate the existence of, a *self-paced* cerebral activity which would be responsible for spontaneous movements and, by extension, nonexternalized (therefore purely subjective)
phenomena, such as thought? The response to these questions that
was sketched out at the end of the last century could only be
theoretical in nature. The fund of neurophysiologic knowledge of
the era, however considerable, did not allow for any sort of demonstration of cerebral activity beyond the observation of movements
themselves; the direct means of recording which we have at our
disposal now were developed much later. Beyond this, the reflex
theory with its sensualist basis, associationism, and the new behaviorism all still supported the idea of cerebral function as strictly
dependent upon external contributions. The brain was certainly
conceived as capable of memorizing a part of the information received, and of transforming it throughout the course of its propa-

gation from center to center, but movement always appeared as the final result of this process, which started at the level of receptors and ended as the restitution, more or less direct and more or less changed, of the initial energy. Such was the conception of Pavlov.

Among the theories which marked the evolution of ideas in this domain, it is fitting to give some attention to that of S. Freud around 1895. His theory, which attempted to describe psychic function in terms of neuronal mechanisms, took reflex theory and associationism as its base but enriched them with a rather specific addition. According to Freud, neurons obeyed a primary function, which he designated as the "principle of inertia." Thanks to a "discharge" mechanism, a neuron gave up the information it received and could maintain itself in a state of nonexcitation. Reflex movement was thus the primary means for neurons to discharge their information. Under certain conditions, however, "the neuronal system becomes obliged to renounce its original tendency towards inertia . . . It is forced to learn how to support a stored quantity, which will be sufficient to satisfy the needs of a specific act. According to the fashion in which it does this, however, the same tendency may persist under a modified form as an effort to maintain the quantity at a level as low as possible."[1]

In opposition to the primary function, which rested upon the simple notion of a "current" traveling through the neuron, the secondary function of retaining a certain quantity of information presupposes the more complex notion of "contact barriers" that resist the passage of this current. Certain neurons, the *phi* neurons, which were used for perception, readily allowed themselves to be traversed without retaining anything; others, the *psi* neurons, resisted and retained much current: "Memory, and probably also the psychic processes in general, depend upon these latter neurons."[2] However, the primary function determined the system's overall functioning. Even if, "under the pressure of the exigencies of life, the neuronal system is forced to accumulate reserves in quantity,"[3] it avoided being filled up by this quantity, that is, being "invested," by establishing sinks across the contact barriers in order to discharge itself.

What was the difference between the *phi* system and the *psi* system based upon? According to Freud, it was most assuredly not morphologic criteria, but rather the respective "distance" of each type of neuron from the periphery. The *phi* neurons were in contact with the exterior of the body: they were constantly being bombarded by vast quantities of information and had to discharge these quantities as quickly as possible. The *psi* neurons (which Freud localized in the "grey matter of the brain"), on the other hand, were without contact with the external world, and only received information from *phi* neurons or from the interior of the body—in any case a much smaller quantity than the *phi* neurons received. They could thus store information in the form of energy without risk of becoming invested too quickly. This reserve of energy which formed in the *psi* neurons permitted an intrapsychic circulation of quantities of information, whose degree of accumulation or discharge explained, respectively, the states of desire or affect. It was thus that the will consisted of "a discharge of the total *psi* quantity,"[4] according to Freud.

The systems described by Freud obviously tapped their energy from sources outside the organism, or at least outside the brain. A small part of this energy was, in effect, also presumed to come from internal sources, derived from sensations of hunger or sexual drive. In this respect, we can note the similarity between Freud's theory and Pavlov's, even if the distance between these two authors becomes enormous if one refers to their respective conceptions of the determination of behavior. For both authors, it was sensations coming from outside or inside which fed the cerebral machine, and caused it to function. Did this mean that a suppression of afferent pathways, which conducted sensations toward the brain, would result in an arrest of its function?

This hypothesis had been seriously considered, in order to interpret the effects produced by the destruction of an organ poorly understood at that time, the labyrinthine apparatus located in the vestibule of the inner ear. In 1825 Flourens ablated the labyrinth in a pigeon and noted that the animal turned on itself and fell, as if vertiginous. He suggested that the labyrinthine apparatus exercised a moderating influence on the brain, and its ablation consequently liberated motor reactions the animal could not control. Later, it was noted that the destruction of the labyrinth in fact abolished muscle tone on the side of the lesion, which was felt to determine the animal's fall. Finally, in 1892, J. R. Ewald synthe-sized these diverse interpretations, and to him belongs the credit for having discovered the correct mechanism, and above all for having furnished, for the first time, a solid physiologic basis for the notion of spontaneous neural activity. If, Ewald thought, the animal turned and fell after the ablation of the labyrinth on one side, it was as much because of the loss of muscle tone on that side as it was because of increased tone on the intact side. His conclusion was therefore that in the normal situation the two labyrinths, in concert, exerted a tonic symmetric influence on the nervous system, thus producing a state of permanent excitation, a nervous "tone" thanks to which muscle tone could be maintained. This hypothesis, now largely verified,[5] certainly explained the regulation of balance and posture, but it also seems to have had profound repercussions on the whole of research regarding spontaneous behavior.

It was basically along these same lines of thought that the German biologists at the beginning of this century, such as J. von Uexküll, interpreted the influence of the level of sensory stimulation on the motor reactions of invertebrates. Thus, an insect reacted more intensely to a stimulus if the setting in which it was tested was more brightly lit.[6] This relationship, thought von Uexküll, was due to the greater or lesser quantities of tonus accumulated in the "reservoirs of tonus" located in certain regions of the nervous sys-tem, and fed by the activity of sensory receptors. In order to verify this theory, it was only necessary to ablate one eye or one antenna of an insect; the animal immediately began to turn round the intact side, as if led to this side by a higher state of stimulation. The sensory organs were thus, to use von Buddenbrock's expression, *Stimulationsorgane*, constituting the very generators of spontaneous neural activity. This solution only displaced the source of neural activity toward the sensory organs, making it once again strictly dependent on the external milieu, and thus maintaining intact the notion of exogenous determination of the whole of behavior.

The sensory organs, which captured the brusque variations of the environment, were also considered sensitive to stable states of the environment. In terms of energy, it was thought that the tran-sition between states in the environment provided the energy for reflex behavior, while the stable states themselves furnished a basal energy, a reserve which could be mobilized during spontaneous behavior. The theory of *Stimulationsorgane* turned out to be incorrect, at least in its original formulation. With the single exception of the labyrinthine receptor, the sensory receptors do not, in fact, register anything except transitions between states in the environment; they rapidly "adapt" when a state of stimulation is maintained. The size of the discharge representing the activity of the receptor increases at the moment of application or cessation of stimulus, and then decreases in several fractions of a second.

The theory which gave the sensory organs a preponderant role in the genesis and maintenance of spontaneous cerebral activity nonetheless survived until very recently, in a slightly different form, as an explanation of the alternation between waking and sleep. After the experiments of Flourens, of which we spoke in the first chapter, the idea that sleep was a "passive" state, characterized in some sense by the absence of attention, gained ground. The pigeon whose cerebral lobes were removed "no longer moved spontaneously, almost always affecting the appearance of a sleeping or somnolent animal."[7] The state of wakefulness was thus defined as positive, and by contrast the state of sleep was defined as negative. The spontaneous activity of sensory receptors seemed to find its place naturally in this explanation: by constantly bombarding the brain, the receptors kept it awake, while inversely, the cessation of this bombardment produced the appearance of sleep. According to N. Kleitman, sleep was due to a sort of functional de-afferentation of the neural centers which normally were responsible for the state of wakefulness. According to this hypothesis, wakefulness was at first (in the newborn) a transitory phenomenon, an awakening "by necessity," conditioned by endogenous stimuli such as hunger. The relative privation, at this age, of stimulation arriving from the outside world explained why the newborn was usually asleep. Later in life, however, the subject remained awake "by choice," consciously maintaining a high level of excitation in its wakefulness center, supplied by sensory receptors; to fall asleep, one needed only to reduce the sensory bombardment by assuming a recumbent position and suppressing visual information through darkness and closing the eyes.[8] Neurophysiologic experiments supported Kleitman's hypothesis. F. Bremer, using the cat as the experimental animal and applying the technique of electroencephalography, then quite new (see below), noted that a rostral section of the brainstem produced high voltage, slow electrical waves across the cerebral cortex, waves which were normally associated with the sleeping state. This section suppressed the access of cutaneous and proprioceptive afferents, as well as auditory and labyrinthine afferents, to the cortex. Hence, sleep, according to Bremer, was the result of the suppression of "the incessant influx of excitation, especially cutaneous and proprioceptive in origin, which are essential to the maintaining of the awake state of the telencephalon, an awake state which the olfactory and optic apparati alone are not apparently sufficient to maintain."[9]

All the conviction that a powerful theory can give was certainly necessary to assimilate normal sleep into the state of an animal with such a mutilation of its nervous system, and to reduce the effects of a section of the brainstem to the mere interruption of sensory pathways. Still, this same reasoning was used by G. Moruzzi and H. W. Magoun in 1949 to localize the "wakefulness center" in the reticular formation of the brainstem. The destruction of this vast ensemble of neurons produced the same effects as rostral sectioning of the brainstem in Bremer's experiment: cortical electroencephalographic activity was taken over by permanent slow waves. The interesting point of the lesion produced by Moruzzi and Magoun was that unlike a complete section, they spared the sensory pathways, which are located outside the reticular formation. The modified theory thus became the following: Sensory afferents did indeed maintain the awake state, but via the intermediary of the reticular formation. The anatomy of this structure, known for a long time, included sensory pathways which, in the course of their traveling to the cortex, gave off collateral branches that innervated the reticular formation. The reticular formation then retransmitted the excitation it had received to the whole of the thalamus and cortex. However, sensory information lost its specificity (visual, somesthetic, etc.) in penetrating the reticular formation. According to Magoun's followers, this sensory information only served to produce a "reticular tone," which conferred the role of a veritable *activating* system upon the reticular formation. The lowering of the reticular tone, whether produced by a reduction in sensory afferents, by cortical inhibition, or even by medication, produced a cascade of de-activation or de-afferentation from thalamus to cortex, with sleep as the result. It is therefore not surprising that the destruction of the reticular formation, even if it spared the direct sensory pathways, deprived the animal of the capacity to put itself in a state of attentiveness, or alertness, and above all deprived it of the capacity to remain in this state of wakefulness which was the necessary condition for the externalization of an action or a behavior.[10]

The conception of spontaneous cerebral activity had nonetheless evolved little with the discovery of the reticular formation. The notion of reticular tone obviously invoked von Uexküll's notion of a "tone reservoir"—the same dependence upon the sensory organs and the environment, the same loss of specificity, with sensory information transforming itself into undifferentiated energy. Moreover, the distinction between two categories of structures in the nervous system—those sensory systems called "specific" and the reticular or "diffuse" system, called "nonspecific"—was reminiscent of the distinction Freud made between *phi* and *psi* neurons.

These various historic associations gave this theory some of its respectability, and favored its rapid growth. Thus it was that between 1950 and 1960 the reticular theory had become a generalized explanation of behavior, explaining sleep and wakefulness, attention and the reaction of orientation, as well as certain mechanisms of learning such as habituation. Subsequently, however, numerous discoveries brought this supremacy into question. It is now known that the reticular formation is not a diffuse system, but that it is differentiated into subsystems, well-organized and involved in well-defined operations. Moreover, notions about the unitary nature of wakefulness, for which the reticular formation represented the neurologic incarnation in some fashion, began to splinter. Each sensory system (visual, auditory) is thought to possess the capacity to control attention in its own sphere, without resorting to some global process where sensory information mingles and loses its specificity. Finally, and above all else, sleep is now considered as an autonomous phenomenon, which begins in an active manner, and no longer as a simple passive state which is only defined in the negative with respect to wakefulness.[11]

One of the first experimental contributions to demonstrate the active character of sleep was that of R. Hess at the beginning of the 1930s. Hess made a distinction in the nervous system between an "animal" system charged with regulating relations with the environment (in particular, the sensory and motor functions) and a "vegetative" system charged with regulating organic functions. He subdivided the vegetative system into two parts: one which dispensed energy (ergotropic) and the other which tended to economize and protect it (trophotropic). The animal and vegetative systems together found their underlying rationale in a teleologic principle of biologic order oriented toward the preservation of the "unity of the organism" and, finally, toward the possibility of man achieving his "vital program."[12] Sleep, which Hess considered as one of the manifestations of the trophotropic vegetative system, thus had to have a positive function in the realization of this biologic order. In 1933, he stated, "this concept leads inevitably to the important consequence that the order is established by an active mechanism."[13] Therefore a sleep "center" must exist since biologic order requires specific neural structures before it can become an observable fact.

By using electrical stimulation techniques in the brains of animals who were otherwise normal and free to move, Hess ended up localizing not only the cerebral zones where stimulation produced sleep but also other zones which seemed to control more or less complex motor behaviors. Unlike electrical stimulation of the cortex, stimulation of limited areas of the brainstem produced coordinated, organized movements, and not contractions of isolated muscles. When stimulated, a cat turned its head and eyes toward what seemed to be a noise or an object; if the stimulation were more intense, it turned and threw its paw up. In other, higher regions (the diencephalon), defensive or attack behaviors were obtained. According to Hess, this was all evidence that the motor system (which he called the "biomotor" system) was organized according to principles which reflected the needs of the organism. His essential argument was that, although the number of combinations of muscles was very high, only certain combinations were observed in the course of a stimulation. "Only the biologically important [combinations] . . . have been developed by nature. This reduction means economy," he stated.[14]

Electrical stimulation of the brain is certainly an experimental method open to criticism when it is used outside of the context of the delineation of motor localization. For many authors, results having a functional significance for the organization and institution of a behavior should not be expected from this technique. Nonetheless, the experiments with stimulation under Hess's influence gave rise to discussions which should be interesting for us in terms of understanding spontaneous motor activity. It was clear that stimulation itself, when applied to zones that were not simply effector, did not produce the muscle contractions which composed the response: rather, they appeared to liberate a pre-established set of contractions, identical to those which might have been produced by some natural cause. In other words, the type of behavior "evoked" by a stimulus depended more upon the structure and organization of the stimulated zone than on the stimulus itself. One could compare the role of the stimulus to that of an operator who, by the simple pressing of a button, started a machine. The complex operations executed by the machine are not "contained" in the movement of the finger which pressed the button, but rather in the manner in which the machine was constructed and in the program which it contained. In fact, after Hess, numerous authors were able, by electrical stimulation of the brains of animals, to obtain organized behaviors such as licking, mastication, swallowing, scratching, attack, vocalization, etc. In humans, the effects of stimulation during therapeutic neurosurgical procedures were observed: depending upon the zone stimulated, the patient made a gesture with the hand as if to brush off his clothing, made mouth movements as if he were eating, etc. Occasionally, when the stimulus was applied to certain points in the thalamus, language activity was produced: the patient pronounced a short phrase like "let's go home" or "I hear something." Repeated application of the stimulus in the same location produced the same effect.[15]

From this work, notions of a particular form of motor organization, located somewhere between the reflex proper and the truly spontaneous act, seemed to emerge. This intermediary form seemed to require a precipitating stimulus in order to be manifest. Unlike the reflex, however, the stimulus was only a "context" here, and controlled neither the intensity nor the duration of the reaction. To the contrary, once begun, the action proceeded in a stereotyped, automatic manner, far surpassing the duration of the stimulus. There were numerous examples of these "precipitated actions," from classic automatism (mastication, locomotion) to complex behavior described by ethologists. The point these diverse manifestations had in common might be the existence of neural generators, functional entities where action was permanently represented, ready to be used. The notion of a precipitated action is certainly much closer to that of spontaneous behavior than is the reflex, even if one includes the psychic reflexes of Setchenov or the conditioned reflexes of Pavlov in the latter. At any rate, it implied a very strong, dependent relationship between the neural substrate and behavior, conceived of here as the addition or juxtaposition of a limited number of defined actions.

Such a neurobiology of action has great difficulty avoiding exploitation by the supporters of the "innate" conception of behavior, as formulated by K. Lorenz and his followers. This conception implied that behavior consisted of chains of actions belonging to each species (their *repertoire*), and that these specific actions were dependent upon the species' genome. Each chain of action was brought into play by a precise stimulus, normally present in the ecologic milieu of the species under consideration; it would come forth of itself in a "spontaneous" manner once triggered. This theory unified in one conceptual whole the notions of a genetic program (the molecules of the genome coding behavior), a motor program (the neural structures containing an encoding of the sequence of actions), and a vital program (the sequence of actions obeying a biologic order). From this viewpoint, "will" could only be one of the terms in a teleologic determinism which, no more than the vitalists of Pflüger, Hess was unable to avoid. The will, said an author such as J. M. Delgado, had as its principal role "to trigger previously established mechanisms." "Voluntary and electrical triggering can activate cerebral mechanisms in a similar way."[16]

Lorenz's thesis regarding this subject might, at first glance, seem paradoxical. According to him, the idea that behavior could be a collection of reactions to the environment and because of this, be modifiable by learning was "entirely false." In fact, "the central nervous system has no need to wait for stimuli before responding, like an electric bell waits for someone to push the button. It can produce its own stimuli."[17] The explanation was as follows: "When an instinctive behavior . . . is stopped for a long time, the threshold for stimuli which trigger it is lowered; . . . the lowering of the threshold for triggering stimuli can, in certain cases, approach zero, that is the instinctive movement in question can "take off" without any external stimulus."[18] The denial by reflexologists, and subsequently by behaviorists, of the existence of long sequences of innate behaviors, and the possibility of their spontaneous triggering had, according to Lorenz, blocked research in this domain for a long time. The mechanistic thinkers who belonged to one or another of these schools had limited themselves to experiments in which a change in the state of the milieu was created, and then the response of the animal to this change was observed. "The *a priori* idea according to which the reflex and the conditioned reflex are the only fundamental elements of animal and human behavior thus determines a type of experimental array . . . in which the nervous system has no occasion, so to speak, to show that it is capable of doing something other than respond to external excitations which work upon it."[19] It was in observing the behavior of animals left to themselves that the spontaneity of their instincts could be seen. Pavlov certainly recognized the existence of relatively complex innate behaviors belonging to each species and oriented in an ultimate way toward the preservation of the species. However, these absolute reflexes (of which we have spoken in Chapter 3) conserved the fundamental charcteristic of reflexes by being triggered by an external stimulus.

Lorenz's explanation of the spontaneity of instinctive behavior could seem deceiving. It was a question, he thought, of the accumulation of "emotivity," "continually secreted by the organism and spent in the course of the movement under consideration."[20] This function of accumulation was an "elementary aptitude of the central nervous system," producing a "spontaneous generation of excitations," the whole constituting a "primitive function hitherto unknown in the central nervous system."[21] In fact, in comparing this property of the nervous system with the cardiac automatism, Lorenz made explicit reference to the theory of nervous *oscillators* developed by von Holst starting around 1930. After observing locomotion in insects and fish, von Holst arrived at the idea that certain neural elements localized in the motor centers oscillated, according to some permanent rhythmic activity: neurons did not merely serve to transmit information from place to place, retaining a greater or lesser quantity; they also fabricated information themselves. Each oscillator had a characteristic period, which could vary to a certain extent. The synchronous oscillation of several elements resulted in a movement, or a series of movements. Finally, small variations in phase between different oscillators could explain the variability of movements when they were forced to adapt themselves to external conditions.[22] There is, at present, scientific proof of the existence of von Holst's oscillators: nerve cells which possess a spontaneous rhythmic activity have been found in invertebrates as well as in the higher mammals. These cells constitute the basic element of veritable biologic *clocks* which are involved in the generation of automatic movements, and also in the secretion of hormones by the nervous system, the regulation of sleep, etc.[23]

These generators, to continue the terminology used above, would be to the regulated unfolding of "automatic" behavior sequences what the mechanisms invented by Vaucanson were for the functioning of automatons. They allowed one to compare the function of the nervous system to that of a machine that "worked by itself" once prompted or, at the extreme, could trigger itself. However, such a physiology of spontaneity gives a singularly limited explanation of behavior. If it explains the automatic aspects quite well, it cannot, by comparison, be generalized to the other aspects of behavior without producing a behavior without a subject. Such is the inevitable result of any finalism, even dressed up as vitalism: in accepting submission to a biologic order, via one of its multiple justifications (the future of the species, the preservation of the individual, the integrity of the organism, etc.), only two alternatives are left. Either the subject becomes nothing but one more justification for this order which supercedes it, or it is relegated, by scruple or precaution, to a sphere apart, outside of biologic rules.

Another aspect of the physiology of spontaneity is the existence of spontaneous cerebral activity. Whatever hypothesis is used to explain the *onset* of an action (the notion of endogenous stimulus being nothing, after all, but a difference in style), it is clear that this onset occurs before the action: muscular phenomena are only the final result of the process. By investigating the brain itself, would it be possible to trace the action back in time, and have access to the basic mechanisms which prepare it? The techniques that permitted investigations beyond the observation of movement, previously the only means of study available to cerebral physiology, had a difficult birth. R. Caton, Sherrington's predecessor in the chair of physiology at Liverpool, was undoubtedly the first to have obtained any results in this area. Researching the cerebral counterparts of the electrical potentials recorded at the level of the nerves, Caton applied electrodes to the surface of a rabbit's brain. He observed the existence of spontaneous potentials, occurring in the absence of any motor activity and independent of cardiac rhythm and respiratory activity. These potentials only disappeared with the death of the animal. This type of work aroused minimal interest in Caton's time, and was not pursued. Not until Hans Berger's publication of 1929 was *electroencephalography* born; it developed rapidly from 1935 onward. The tracings obtained in humans by Berger did not, at first, convince his contemporaries: the man was considered an amateur, a bit native, and the technology was primitive and not very reliable. It was necessary to await the "rediscovery" of the phenomenon by E. Adrian at Cambridge for it to be universally accepted.[24]

Berger had ambitions other than simple technical performance: he saw in the recording of cerebral electricity the means of approaching more closely the "psychic energy," the point of departure for mental phenomena. The brain, he thought, was the site of metabolic activity like no other living tissue. This activity liberated a chemical energy which was subsequently transformed into heat, electricity, and psychic energy. It was, therefore, logical to infer the production of psychic energy from the production of electricity. It was not until 1965, however, that the problem of the electrical translation of psychic energy into spontaneous movements was tackled. From this date on, the potentials which could be translated as motor preparation were recordable.[25] Thanks to the use of computers, the appearance of a slow variation of the potential, preceding the movement by 1 or 2 seconds and increasing in size right up to the execution proper, can be observed. The more complex the movement, the greater was the delay in its appearance. This electrical phenomenon is spread in a relatively larger manner bilaterally across the surface of the skull, except during the tenth of a second which precedes the execution, when it is predominantly localized over the cortical motor zone corresponding to the movement (for example, over the inferior part of the left motor cortex when the right hand moves). This type of potential can also be observed before the appearance of a movement executed as a response to an external signal. However, in this case, the variation in the potential begins immediately after the signal, and evolves much more quickly toward the execution of the movement, which lasts around one-third of a second. As Eccles noted, "The voluntary command, occurring in the absence of any initiating stimulus, is very weak and requires hundreds of milliseconds in order to accumulate a cerebral excitation intense enough to give rise to a motor discharge."[26]

If the recordings furnished by this technique are to be believed, the voluntary activity is prepared, or better still, *constructed* by a progressive process. The decision, the selection of the goal, the organization of the entirety of a movement take time, and require the association and then simultaneous activation of a great number of neural "modules." We are thus entering the domain of mental representation and its transformation into action.

8

Representation, Plan, Program

*I*n the domain of the generation of movements, how far can the claims of the postulate of objectivity be taken? Can it also explain the psychic step, during whose course an action is decided upon and then prepared for? Few were bold enough to cross the implicit limit traced by observation, the majority preferring to invoke some *deus ex machina* to explain our representation of the world, our thoughts, and our actions at the least cost. Thus Descartes imagined a *homunculus* placed in the interior of the subject, who could, at its leisure, inspect the images of objects impressed on the retina and brain. The last century saw numerous satirical cartoons featuring a little laboratory in the interior of the skull, where a scientist was more or less facetiously playing with vials and apparati. In the end, the cerebral machine of the nineteenth century was no more intelligent than the automaton of the eighteenth century.

Bergson's well-known thesis regarding the relationships between perception and action would seem to serve as a good introduction to this chapter, since the manner in which the problem was posed so clearly reflects the context of the time. For Bergson, "the brain cannot be anything other than . . . a sort of central telephone bureau: its role is to 'send on communication', or make it wait. It adds nothing to what it receives; but since all the perceptive organs send their last extensions there, and since all the motor mechanisms of the cord and the brainstem have their specific representations there, the brain actually constitutes a center where peripheral excitation can he put in contact with one or another motor mechanism, which is chosen and no longer imposed."[1] Thus, for Bergson, the nervous system plays the role of a simple conductor, connecting the periphery to the center and the center to the periphery. The environment can pose "questions" to motor activity via this conductor: "Each question posed is what we call a perception."[2] Perception is different from the image of the object, in that it is not located in the sensory neural elements, nor in the motor centers. It does not arise from a representation of the outside world assembled piece by piece, but it appears where there is an action, so that "there is no perception which does not extend itself

Figure 13 "The brain is the laboratory where all the phenomena of intelligence, all the psychic transactions that make us an occasionally reasonable being, take place." Drawing taken from *La grande nevrose*, by Dr. J. Gerard (Paris, 1889).

into a movement."[3] The brain is an instrument of action, and action alone; it is neither the site nor the explanation of some representation. Moreover, Bergson said, there is "only a difference of degree, and not of nature, between the faculties of the brain we call perceptive and the reflex functions of the spinal cord. While the spinal cord transforms the stimuli received into movements more or less executed of necessity, the brain puts them in contact with motor mechanisms more or less freely chosen." What the mechanisms of the brain allow us to explain is "our actions, whether begun, prepared, or suggested; they are our perceptions themselves."[4] Perception corresponds, in fact, to the cerebral state which accompanies the transformation of sensation into action, some sort of vital property of sensorimotor activity, what others would call an "emergent" property.[5]

Even the idea of a mental representation, and an elaborate treatment of sensorial information by the brain, was not accepted by everyone. For the behaviorists—and in particular J. J. Gibson, their representative in this domain—perception was "direct" and, as a result, implied neither a representation of the outside world nor treatment of information. "Stimulation is a function of the environment, and perception is a function of stimulation."[6] This attitude presupposed that the environment was organized and specified in an "ecologic" manner, and that the critical stimuli present in this environment were invariant, and were grouped as sets of information. Thus, for Gibson, the simple fact that the subject was displaced with respect to these sets produced a continuous transformation in their appearance, very different from that which was produced when the environment was displaced with respect to the subject. From this fact, it was no longer necessary to imagine a particular mechanism of the type that were envisaged in Chapter 6 to explain the difference between ex-afference and re-afference. Perception was nothing other than the direct and immediate acquisition of information: "No mental operation on units of consciousness, nor a central nervous operation on the signals in nerves,"[7] was necessary. In extending this reasoning to the extreme, one could even imagine the hypothesis of a *selection* (in the evolutionary sense of this word) by the environment of the neural mechanisms of information capture and the best-adapted behavioral responses. As for mental states (how could one deny their existence?), they did not intervene in the stimulus–response cycle which controlled behavior. In 1890 William James had already considered feelings and other internal states as by-products of our behavioral responses: I am afraid because I flee, I am happy because I laugh (and not the reverse).[8]

Currently, however, behaviorism has been refuted in all aspects, not through a return to some neovitalism nor to a psychology of introspection (though these two tendencies still exist), but rather through an attempt at an objective approach to mental states themselves. Some spectacular experiments have been performed in this domain, particularly those of R. Shepard in California in the early 1970s. A subject was shown the image of three-dimensional objects projected in pairs on a screen for a very brief period of time. The objects were geometric forms of varying complexity. In certain cases the two objects of the pair were the same, distinguished only by the angle at which they were presented. In other cases, the two objects were presented at different angles but were also different in shape, one being the mirror image of the other. The subject had to determine, as quickly as possible, whether the objects were the same or different; the time taken to respond was measured precisely. Shepard demonstrated that the time varied as a function of the angle at which the objects were turned with respect to each other, the response being more delayed as the angle was increased. It should be noted that the duration of presentation on the screen was very brief (several fractions of a second), and the subject did not have the objects in sight when he prepared his answer; it was therefore his internal representations of the objects that he compared. Shepard's conclusion was that the subject's comparison was achieved by the mental "rotation" of one of the objects represented (in such a way to bring it into the same "position" as the other), and that this rotation took more time when the angle which separated the two objects was larger. In other words, the mental representations *were manipulated like objects*, and this manipulation took time. One can verify this easily by asking someone to represent a familiar itinerary with their eyes closed, and then mentally travel it. One interrupts him in order to ask "where" he is: the distance traveled will be greater the longer one waits. In the same way, mentally "writing" a long word requires more time than "writing" a short word.

Ever since Wundt and his school, psychology has traditionally considered time as a representative dimension of the mental processes. The length of time taken by a subject to respond to a determined modification of the external world (the reaction time) varies to an important degree as a function of the situation. In certain cases, the length of the reaction time simply expresses the level of attention or the degree of the subject's preparation: for example, it becomes shorter when the stimulus to which the subject must respond is preceded several seconds by a preparatory signal. In other experiments, however, the duration of the reaction time reflects the complexity of the task to be accomplished in response to the stimulus, especially if the situation includes a choice between several possible options, or any uncertainty about the type of response that should be furnished. Let us take an example: a subject holds his finger down on a button placed in front of him. Periodically, a light appears, either to the left or the right of the button, at varying distances. The subject's task consists in rapidly moving his finger from the button to the light. The time between target illumination and button release constitutes the reaction time. In one experiment, the target which is illuminated is always the same, and consequently can be reached by a repetition of the same gesture; reaction time is very short. The experiment is then complicated by the introduction of variability in the target presentation, appearing either to the right or to the left of the button. Depending upon which side the target appears, the gesture produced is different, since in one case extension at the elbow is required and in the other a flexion is required; reaction time becomes longer. It becomes even longer still if there is variability in both target location and target distance with respect to the button.[9]

In order to interpret the results of his experiment Shepard postulated a mental representation of the *image* of the object, whereas the reaction-time experiment implies the existence of a representation of the *gesture* to be made in order to achieve a determined goal. Taken together, these two experiments support the idea, in two very different situations, of the *indirect* character of the relationships between the subject and the outside world. The subject made his own representation of the world, and this representation guided his action. From this perspective, action was not a direct response to an external situation but rather the consequence or the product of a representation.

The problem could be posed in an even more radical fashion, as, for example, Ulric Neisser did in his book *Cognition and Reality*.[10] Perception and action were linked in a cycle: perception was as much the result of information filtered and treated by the sensory organs as it was the active exploration of the environment. However, according to Neisser, this exploration does not occur by chance; it is guided by a "schema," a plan of action which anticipated, in some fashion, reality. In one sense, the representation of the world preceded perception, perception being placed in the context of a "cognitive map," a sort of mental apparatus which imposed structure on the environment. In the perceptive cycle of Neisser, even if external reality acted on the schema (by correcting it or bringing it into being), the schema was nonetheless "already there" before the arrival of information. Such a nativist approach to behavior, several examples of which we have already mentioned, obviously takes its full significance only when it is applied to the problem of ontogenetic development.

Neisser's approach was oriented toward the perceptive aspect of representation: action does not intervene in his cycle except when it participates in the perceptive cycle, and allows for the confirmation of the predictions made in the context of the cognitive map. Is it possible to apply this reasoning to the description of a motor cycle which, itself, constituted a schema anticipating the execution of movements? As Neisser had predicted for perception, the motor schema has to consist of several levels embedded in one another, each level being defined by its degree of generality or complexity, thus by its content. We will use a hierarchial classification of the different levels of representation of action. However, we must be very clear that, here, the concept of hierarchy owes nothing to the hierarchies of association or connection. When information circulates from one level to another, it circulates in two directions, that is, it circulates just as much from the lower to the higher as from the higher to the lower level. The hierarchy here only implies a difference in competence (the higher levels have general competence and therefore poor specialization, while the opposite is true for the lower levels, which are very specialized), and not a dependence on one-way traffic (the lower levels do not necessarily "obey" those levels located above them).[11]

In the definition of the higher levels of motor representation, terms such as *intention* or *goal-directed action* take on a particular importance. To say that a movement is intentional is to imply that it is preceded by a representation of the effect that it would produce once executed. In other words, intention is the representation of a goal to be attained or a certain effect to be produced. When the goal is attained and the effect is produced, that is, when the intention is "achieved," the representation of the goal and the attained goal become one. The term *intentional action* is therefore very different from the term *voluntary action*, although the two share some aspects. Whereas the term *voluntary action* strongly implies the notion of an agent, or a cause, of the action, the term *intentional action* implies, in addition, the notion of a final state (a composite of the subject and its environment) which is different from the initial state. This difference is precisely the intention, or better still, as we have already said, the representation of the goal. It is therefore logical that the more explicit term *goal-directed action* is now preferred by many to those vaguer terms (which included a metaphysical connotation) of intentional action and, *a fortiori*, of voluntary action.

The notion of goal-directed action inevitably evokes the idea of a "searching" machine, somewhat similar to a guided missile. It is true that a guided missile has in its interior, by its very construction, a representation (albeit rather crude) of the goal to be attained. It can therefore direct its own action as a function of this goal, eventually correcting its trajectory according to the messages of its "sensory" apparati, until the final state which corresponds to the representation of the goal is attained. This type of mechanism, well known to cyberneticists, is undeniably valuable in explaining certain aspects of movement control in living subjects. It does not, however, really take into account the multiplicity, nor above all the complexity, of the factors which come into play in the constitution of a goal. One might moreover object that, in certain cases, the goal toward which a person directs its action can be very vague or even nonexistent, and that, as a result, the final state is never reached; a missile, on the other hand, cannot satisfy itself with an imaginary goal. Could one consider as a goal the simple target, which immediately calls forth, and in some sense already contains, the only action which can be exercised in its direction? This distinction allows one to begin to deal with the problem of the content in the different levels of motor representation.

The highest level could be considered a "conceptual" level, which contains the plan of action and its different phases, as well as a certain anticipated evaluation of its consequences. But this representation of the goal of the action, and the means of attaining it, must also contain information on the objects toward which the action is directed, and on the final result of the transformation it will make them eventually undergo. Only in this fashion can a global representation of the interaction between the subject and its environment be established—a problem posed by any action directed toward a goal. There is no border between the perceptual domain and the motor domain of our representation of the outside world, as Bergson had thought. Action directed toward an object confers on the object particular qualities which become constituent elements of the program of this action. The fact that an object is manipulable, that it has an instrumental function (by design or by accident), or the fact that it is an obligate relay for the pursuit of the action (such as a book, flint and steel), or that it is transformable or even edible, represents what Gibson called its *affordance*. The qualities which define the affordance of an object define (at least in part) the form of action which is going to be directed toward it.

The perceptual *and* motor character of the representation of an action pose some interesting problems in detail when we study the level on which the mechanisms proper of this action operate. Before we can do this, it is necessary to try to give ourselves an idea, however approximate, of the functioning of the conceptual level, as well as the possible consequences of its malfunction. Let us take the case of someone who, remembering that his car is incorrectly parked, thinks, "I'm going to park the car." This simple proposition defines a complex sequence of movements, which should normally terminate in a precise goal: parking the car alongside a curb. These movements must be executed in a certain order: find the ignition key, walk to the car, get in, start it up, etc. The sequence anticipates the final goal, but also anticipates each of the partial goals which occur in the course of the action. The fact that we are absorbed by looking for the ignition key we have momentarily lost does not turn us away from the final goal, which includes this search, since the goal cannot be attained in any other way. In other words, the proposition "I'm going to park the car" constitutes a global representation oriented toward a goal in the long run, but which includes much more than the announcement of the goal. This representation must also include the risks that the current position of the car entails, its functioning, the ignition key and its possible hiding places. Everything seems to happen as if the subject who wanted to park his car was proceeding with a semantic analysis of the verb "to park." In the sentence, the verb constitutes the node of the action: toward it converge the subject (I) and the object (car). The latter determines the modalities of action, with each verb (finding the key, starting up, backing up) constituting a new node in its turn. Each action can thus be made the object of a pragmatic analysis, "generative" in the sense that, like generative linguistics, its elementary constituents (actors, instruments, circumstances) can constitute as many points of view for understanding its unfolding.

This comparison of the content of the conceptual level to a verbal proposition (I'm going to park the car) illustrates the relationship between human language and action. Human action, when it is voluntary, intentional, or propositional, has the distinction of being dominated by language, in the sense that language precedes the action and is an integral part of the representation. Neurologists have often pointed out the close relationships between certain forms of aphasia and certain forms of apraxia: the lesions responsible for both are usually located in the same hemisphere, the left hemisphere. A recent theory considers this hemisphere to be a specialized structure for the production of sequenced activities, such as language and movement.

Action, however, includes other aspects which are certainly not reducible to an organigram, of the type constructed by the linguists. The representation of an action can also lead to a global imagery where, a bit in the manner of what one experiences in a dream, one can see oneself in the midst of action, where actors and circumstances are quickly apprehended in the complexity of their relationships. To simplify greatly, one might attempt to oppose this mode of spatial or "photographic" representation of an action to the temporal and sequential mode which we invoked above. In pushing this comparison a bit further, one could end up attributing a special role in the construction of this mode of spatial representation to the other hemisphere, the right hemisphere. It is known that lesions of the right hemisphere do not produce disturbances of language, but rather produce a disorganization of "spatial thought," which is translated into a disturbance of one's ability—or even into a total inability—to imagine locations to represent them to oneself, to locate oneself on a map, to construct a drawing, especially if it involves perspective. Language-thought and action-thought on the left, spatial-thought on the right? Such is the simplistic schema that certain of Sperry's experiments might support, experiments which involved patients with sections of the corpus callosum; more recent experiments, in normal subjects or those with lesions, which attempt to dissociate the functions of each hemisphere, might also support this schema. In fact, however, a motor representation is verbal and spatial *at the same time*. It takes into account information processed by both hemispheres, and nothing allows it to be thought of as located more on one side than the other, nor does normal functioning imply some association with specialized centers.

In the description just given, we can see what the consequences of a disorganization of the motor representation at the conceptual level would be. It is not, however, perhaps the right moment to ask ourselves what type of apraxia this disorganization would produce, having repudiated the associationist schema. Dejerine certainly called ideational apraxia an apraxia of "conception," but in this instance, too close a relationship between a given level of motor functioning and a clinical form of disorganization of action would, in itself, be hazardous. Only the inverse of this process (looking for the negative image of normal behavior while observing apraxic behavior, as we did in Chapter 5) seems legitimate. In this case, a cerebral lesion, either revealed at autopsy or by modern techniques of neurologic investigation, would readily allow us to view the motor problem in terms of localization: while unable to reveal the localization of the mechanism normally responsible for a level of function, clinical–anatomic correlation will at least localize the area which produced the observed symptoms. The problem is posed in a very different manner when, as in this case, one starts from a theoretical model whose structure, hierarchy, and content are invented.

The only reasonable hypothesis in this domain seems to be the theory which would have the left hemisphere play a role in the organization of the sequence of the different phases of an action. Basically, when an apraxic patient who has suffered a left hemisphere lesion is presented a series of drawings representing the steps of a daily activity (using a telephone, preparing tea), gross errors are committed—for example, placing the picture "speaking on the telephone" before the picture "looking for a number in the phonebook." In other words, in such patients action seems disorganized in its very conception, beyond any implementation as an act. It is not surprising that the execution itself is just as disturbed when the patient is asked to reproduce a sequence of manual gestures, imitating an examiner, or to perform a task consisting of a succession of operations following a pre-established plan. Although the lesion is only on one side (the left), this particular type of motor deficit affects both hands. Thus, we must acknowledge the existence of a real dominance of the left hemisphere for the conception of action, in the same way that this hemisphere is dominant for language. This property would correlate well with the hypothesis that the left hemisphere is specialized for sequential activities. Thus, even if one does not expect to confine motor representation to a circumscribed cerebral localization, the left hemisphere remains a key structure for the comprehension of motor representation.

At another level, motor representation becomes "operational." The plan of action elaborated at the conceptual level becomes the object of more or less detailed programming. As we have said, the conceptual level has no order to give to the operational level: the hierarchy of levels we are discussing should be thought of only in terms of the information they contain. What the conceptual level transmits to the operational levels concerns the goal to be attained, the general climate in which the action will take place, the nature of the programs to be constructed. In return, the operational level informs the conceptual level of the degree of execution of the action: it is by this constant flux of information, circulating in both directions, that a level can know its own state and influence the state of other levels.

This information exchange can be thought of by analogy with the functioning of computers. Recent theories of artificial intelligence tend to think of the mental states as nothing other than complex programs, of the sort that are applied in computers, particularly when one considers "self-evolving" programs.[12] The implications of these theories go well beyond simple analogy: the program, in effect, has no structural relationship with the computer which executes it, but rather it is only a *functional state* of the machine at a given instant. A mental state like the representation of an action could therefore also be merely a momentary functional state of the brain, without a causal relationship with the way in which the brain is made, or even without an awareness it is the brain which is involved in the first place. This analogy can be pushed too far: could a computer convince me that it was thirsty, and would I actually deduce that it was really thirsty? The brain, its function, and "its" mental representations all proceed from the same biologic entity, the organism; they are interconnected in the core of this organism. It is in this way that it is appropriate to imagine the bonds of causality between different levels of the motor program, between representation and operation. These bonds necessarily imply the existence of a *continuity* from one level to another.[13]

The contents of the operational level can be represented as a set of instructions, whose systematic unfolding results in a sequence of motor orders and, from this, the production of an effect. The operational level must therefore be able to assemble, and eventually conserve in its memory (in the manner of the discs and drums of an automaton, or the perforated cards and magnetic tape of computers), multiple programs (the word is taken here in its simplest sense), which are destined to respond at the command of the conceptual level. The very diversity of the movements to be brought about in order to achieve a set goal clearly indicates that the operational level cannot be a unique structure: it can only be a *distributed* structure, largely divided among the motor zones of the cortex, the diencephalon, and the mesencephalon, perhaps even the cord.

Let us take, for example, the case of a prehensile movement, intended to grasp a book on the shelf. This prehensile movement could be situated in time between the book-searching behavior (eye and head movements) and use behavior (opening, leafing through, reading). The set consisting of the search and manipulation of the book could be included in an action still more complex, such as looking for a bibliographic reference. To isolate the prehensile movement is not, however, artificial since this is what the operational level does each time it executes one of the subroutines which lead to the overall plan elaborated by the conceptual level. To program a prehensile movement is not, however, simple: the arm and the forearm must bend in the direction of the chosen book, and lead the hand to the correct location, within a few millimeters. The forearm must, at the same time, pivot in order to position the wrist in the vertical plane, while the fingers must take the form of a pincer in order to grab the book. This routine, which takes place in less than a second, necessitates an anticipatory evaluation of the location of the book in space, its distance from the subject, its orientation with respect to the vertical, its thickness, and finally its weight.

These diverse properties, certain of which belong to the object itself, such as form and size, others of which are a function of its location with respect to the subject, such as distance, are elements of its *affordance*. Each of them, however, call into play a particular mode of information processing by sensory apparati, as we have already mentioned in Chapter 5; even at the core of the visual modality, it is not the same groups of neurons which process information relative to form or size as process information relative to distance. This consideration thus joins those concerning the *distributed* nature of the operational level. One can easily imagine that this level consists of different sensorimotor channels, each independent, each characterized by the type of sensory information it is capable of processing and by the type of movement it is capable of producing. The sensorimotor channel which processes "book thickness" information produces a movement that is adapted to this information, namely, a contraction of the muscles which form a pincer grasp; the channel which processes the "distance with respect to the body" information produces movements which extend the arm, etc. In order to produce the correct movement to grasp a book, the different channels involved must be activated simultaneously. Simple observation reveals that, in fact, the pincer grasp is forming at the same time as the arm is being extended, and that the fingers do not wait until they are close to the book before adopting the correct posture. The channels therefore function in a parallel fashion, each by itself, one might say; this is possible because of their respective independence—an advantageous arrangement when a large quantity of information must be processed in a minimum amount of time, as is the case here.

The problems posed by the function of what we call the operational level of motor representation were first approached in 1935 by the Russian N. Bernstein, in a surprisingly modern fashion: "There exist in the central nervous system exact formulae of movement, or their engrams . . . We may affirm that at the moment when the movement began, there was already a complete collection of the engrams which were necessary for the movement to be carried on to its conclusion."[14] These engrams correspond quite well with what we have called "programs," since Bernstein placed the "temporal structure" responsible for the unfolding of a movement there. When it is a question of coordinating several muscle groups, several segments, even several extremities (as in the case of prehension that we took for an example), different temporal structures could be envisaged. First, it might be a matter of serialization, where the activation of movement in one segment is stimulated by the movement of another segment. Thus, beginning with a single initial impulsion, a succession of contractions would produce a complex movement, and the coordination of the movements of different segments would occur automatically. This hypothesis, of manifestly reflex inspiration, was supported in its time by the behaviorist school.[15] The other possible temporal structure, according to Bernstein, was that which occurred by simultaneous, or at any rate parallel, activation of movements in different segments. It should be noted that the second hypothesis is quite compatible with the notion of an engram or program, while the first does not require a program.

Despite its advantages, the concept of parallel activation of independent channels poses the problem of coordination between the different segments. Bernstein resolved this by placing another engram, the *ecphorator*, next to the engrams containing the "formula" of a movement; he illustrated the role of the former by an example. He compared the program proper to the grooves on a phonograph record, and assigned the *ecphorator* the role of the phonograph's motor, which was necessary in order to play the record. Other analogies are certainly possible, including that of a sort of temporal filter through which the commands from all the channels have to pass, and which synchronizes the set of commands destined for the same gesture by holding back those which arrived early. In terms of economy of mechanisms, the synchronization of diverse components is probably the simplest way to produce the coordination of a gesture.

We still have to determine the precise content of the representation of the action at the operational level, in other words, we must attempt to define how far the complexity of the programming of movements can go. Is it the force applied to each muscle which is represented? Is braking necessary to stop the movement, or will it terminate on its own by exhaustion of the initial impulse's force? Does the form of the trajectory of the moving segment, which depends on the degree of synchrony and respective force of contraction of the different muscles, have to be programmed, or does it come about on its own? Bernstein remarked that the programming of a movement by the nervous system probably does not "resemble" this movement very closely. Basically, any movement occurs against the sum of forces to which it is submitted (gravity, Coriolis forces), and against the forces created by the load on the extremity (if it is carrying a tool, for example). The existence of this field of forces conditions the initial form of the movement, in particular by distributing the observed effects of the motor orders over time. Orders sent simultaneously to the muscles of the fingers and the muscles of the shoulder will produce a movement of the fingers, then the arm, by the simple fact of the difference in degrees of inertia in the segments to be mobilized. This observation underlines an important point: should factors such as form, inertia, and the initial position of the moving segment be included in the motor representation as well?

The current tendency is not to attempt to conceive of a program of movements in all their complexity, nor a representation of all the degrees of freedom possible to the motor apparatus. For certain authors, the program can be limited to a minimum of instructions, the final form of the movement being obtained by the interaction of these instructions with the set of external forces acting upon the organism. For example, for a given movement it is sufficient for the final position of the extremity (or segment of the extremity) in question to be represented. A corresponding tension would be applied to the muscles involved, and the movement would stop of its own at the correct location, defined in some way by the point of equilibrium between the tension of the various muscles. An example of this mechanism can be seen in the execution of a movement including a complicated trajectory (going round one object in order to grasp another): a precise recording of the trajectory would show a segmentation into as many elementary units of movement as there were curves to be effected, as if the programming of the whole occurred by a sequence of defined spatial positions. Such a hypothesis was used to explain the production of speech: in this case, each phoneme could be represented as a given form of the vocal apparatus, responsible for the production of sounds. In the same manner as for the final position of an extremity, the form of the vocal apparatus for a given phoneme would correspond to a final point of equilibrium between the different vocal muscles, an equilibrium which could therefore be attained regardless of the initial form. The hypothesis of a representation of the final position of a moving segment has the advantage that it does not imply the necessity of taking the initial position of the segment into account, which considerably simplifies the programming of complex sequences, such as those which constitute language.[16]

The origin of this theory can be found rather far back. In 1917 Karl Lashley observed a patient whose lower extremities had lost sensation as the result of a gunshot wound to the spinal cord. Despite the complete absence of sensations from this extremity, the patient was capable of bending his knee at a given angle, or placing his foot at a height indicated by the observer, all without benefit of vision as well. Independent of any sensory control, the movements thus had to depend upon a determination of the final position by the central nervous system. Moreover, Lashley noted that a great number of our movements are executed too rapidly to have any sensory control intervene. This observation eliminated, in the same blow, the theory of reflex sequencing of movements for the execution of motor sequences. During the playing of a musical instrument, finger alternations can, in certain instances, attain the frequency of 16 per second. In the spirit of Lashley, this strongly implies that the succession of these movements must be "coded" before they are executed. It is interesting to push Lashley's arguments regarding visual control of movements, the best known modality of motor control. Since the 1960s it has been known that several tenths of a second must pass before visual information can have any influence on the execution of a movement. If we consider that, in acts of daily life such as extending the arm to grasp an object, the hand is displaced at nearly 50 cm/sec, and that a delay of one-third of a second exists between the observation of an error and the possibility of correcting it, then 15 cm will have been traveled before a corresponding modification of the trajectory could intervene. The length of time required by transmission and processing in the visual pathways, and then the time of neuromuscular coupling, does not allow for any "on line" guidance of our movements by vision.

This observation would seem to prove that movements are executed in the absence of any visual control. The acquisition of a single unit of information would take place at the moment of executing the movement, during the preparatory period which precedes it. The movement would then be directed blindly toward its anticipated goal, in the manner of a ball launched toward a point in space (ballistic movement), with the advantage over a projectile that the moving segment remains attached to the other parts of the body, and can therefore benefit from a precise determination of its final position, as we saw earlier. From the point of view of simple efficiency, however, it is difficult to conceive of the execution of movements requiring great precision without some form of control over the trajectory. Such control could be effected in at least two ways. One could, for example, imagine an intermittent control, with visual information being picked up before the onset of the movement, and then being briefly "sampled" in the course of the trajectory. If they were effected early enough, the samplings could furnish indications on the respective positions of the hand and the goal to be attained, and give rise to corrections in the trajectory which would assure the movement's precision. One could just as easily conceive of the visual control of a movement as a form of more or less continuous guidance of the trajectory by the initial

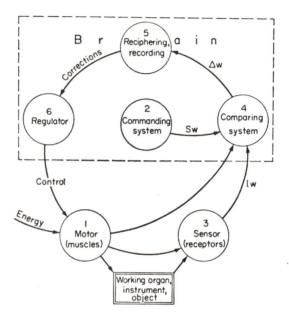

Figure 14 Model explaining the sensory control of movement. according to Bernstein. The control system is based upon the represented values— on the one hand, the goal to be attained (or the "order," *Sw*), and, on the other hand, the actual position of the extremity being moved in the direction of the goal (*Lw*). The comparison between the two allows the separation (Δw) between the order and the execution to be determined. Corrections are produced as a function of Δw, modifying the contractile state of the muscles and complementing the initial expenditure of energy. In its turn, the movement produced by the muscles modified the state of the sensory receptors (visual, muscular), which then furnish a new value of *Iw* for the comparing system, etc.

It should be noted that the value *Sw* is given to the comparing system by the commanding system, placed in the center of the diagram, which also produces the movement's energy. Thus, a "copy" of the motor program is produced which corresponds, for the comparing system, to the goal to be achieved.

information alone, the information available prior to the onset of this movement. It is here that motor and perceptual representations would be joined: the goal to be attained, represented at the perceptual level, would be necessary for the motor representation, in order to determine the final position of the moving segment. At each level of the motor representation (conceptual, operational), an aliquot of visual information would be produced, by what we called the sensorimotor channels (in this case, the visual-motor channels). This "internalized" visual information could thus be transformed into a "reference," a final ideal state. During the execution of the movement, other, nonvisual, information (arising from muscle and joint sensors) would inform the program as to the difference between the unfolding of the movement and the reference: the program ceases to act when the reference is attained.

In a preceding chapter, we have already described a mechanism of the same sort, which, although relatively independent of sensory information, made a comparison between a copy of the efference and the re-afferent information. These concepts, based on feedback control of movement, or on feed-forward control, were inherited from cybernetics. They still represent the best compromise between the extreme positions which postulate a motor activity that is entirely programmed internally, on the one hand, and the opposite, a motor activity that is completely controlled by external forces. The German F. Lincke thoroughly understood this in 1879, during the era when the reflex theory still reigned supreme: "The activity we use to direct our human machine towards its goal results from the difference between the will and observed or imagined reality, that is, the difference between intention and the result of execution."[17]

9

Action, the Force of Self-Organization

Action, when it is directed toward the sensory world, is part of a sensorimotor cycle. It modifies the spatial relationships between the subject and external objects, and the relationships between objects; it also manipulates and transforms the structure and appearance of these objects. The active subject is a party participating in a global interaction with the external world, where perception, action, and new perception follow one another. If action is often a *response* to the environment, it can just as well be a *question* to it. Assessing the adequacy of an action with respect to its goal, or the difference between a motor representation (a subjective hypothesis about the world of objects) and the reality effectively experienced at the moment of action, is the source of learning, of adaptation—in short, of knowledge.

Action is thus the means of verifying whether a hypothesis conforms to reality. In this cybernetic perspective, one can imagine that, in the course of an action, sensory systems behave like error or separation detectors. If a movement achieves its goal, no error is detected, which signifies that the hypothesis regarding the goal conformed to reality. If, on the other hand, some separation persists at the end of the execution, the hypothesis was incorrect. The detection of this separation (its amplitude and its sign) would allow a modification of the hypothesis and thus, little by little, the realization of a movement perfectly adapted to the goal.

In 1961 R. Held saw in this theory the basis for a generalized conception of neural mechanisms for learning. Taking up the notions of efference, re-afference, and efference copy (or corollary discharge), as we explained them in Chapter 6, Held proposed that the program of each movement, at the moment of execution, leaves a trace in some sort of *memory*. He also proposed that re-afference, brought about by the execution of this same movement, was combined in this memory with the corresponding trace. If, as was most often the case, the trace and the re-afference coincided, the combination was saved in the memory, where it grew to be a greatly redundant store of information which stabilized sensorimotor coordination. If, on the other hand, the comparison between the two revealed a discrepancy, the movement program would be rectified, with a view toward a more precise ultimate execution. This model thus left open the possibility of establishing a new sensorimotor coordination in the case where a new combination would take the place of the preceding one, the conditions of execution having been modified in a systematic fashion.

One of the conditions essential for the functioning of this model is that the movement producing the re-afference be an active movement or, in order to preserve the term we have used up to now, a voluntary movement. Movements of this sort are really the only ones which reflect the existence of a program to which the result of the movement can be compared. Held's model demonstrates the renewal of learning theory during the 1950s. Its author certainly avoided a concept where the nervous system, guided by some internal "intelligence," looked for a "certain type of order"; in other words, he avoided the temptation to introduce a teleologic aspect.[1] It is, however, interesting to see to what extent one can avoid this question, above all in the application of this model to the mechanisms of development. Basically, the reintroduction of the idea of active movement into psychology—even if it is a sort of psychology which approaches the neurosciences—inevitably requires that the subject and its internal state be taken into account. The source of action moves from the exterior of the subject, where theories based upon the reflex had maintained it, toward the interior, and the problem is posed of knowing by what principle of organization action is governed.[2]

The experimental basis of this theory rests upon two types of convergent evidence, observed in the adult human on the one hand and in the developing animal on the other. In humans, most of the evidence is taken from experiments which, as Held put it, "re-arrange" the relationship between visual information and infor-mation received in the course of movement. Normally, of course, we sense the object touching our hand exactly at the place where we see it; this is the very expression of sensorimotor coordination. It is very simple, however, to put these two types of information in conflict, by placing an optical system in front of the eyes which changes the direction of light rays to a greater or lesser extent. In this case, the position of objects that one sees through these lenses appears displaced with respect to their true position. One can be convinced of this merely by reaching out for one of these objects.

In his work of 1886, Helmholtz reported this phenomenon: "Take two glass prisms with refracting angles of about 16 or 18 and place them in a spectacle frame with their edges both turned toward the left. As seen through these glasses, the objects in the field of view will apparently be shifted to the left of their real positions. At first, without bringing the hand into the field, look closely at some definite object within reach; then close the eyes, and try to touch the object with the forefinger. The usual result will be to miss it by thrusting the hand too far to the left." A new coordination is not long in being established, however. When one places the hand in the visual field, and one touches objects several times under the guidance of the eye, it can be seen that when the experiment with eyes closed is again tried, "now we do not miss the objects, but feel for them correctly."[3] This refound precision is truly the result of a new coordination since if the prisms are removed and one tries to grasp the object, the hand again misses, but this time by going too far to the right. The persistence of this "after-effect" (the hand going too far to the right) after removal of the prisms represents the recombination, in the memory postulated by Held's model, of the indices regarding "observed" direction and those regarding the "sensed" direction of visual objects. A new coordination had thus supplanted the first, which had been established under normal visual conditions.

Before Held had given these results their new dimension, several observers had tried experimenting on themselves, investigating the effects of this transformation of the visual world. George Stratton, the principal member of this group, asked a question around 1896 which was, in fact, left over from the preceding century, even the seventeenth century: Which sense, vision or touch, instructed the other? G. Berkeley had concluded that touch was pre-eminent,

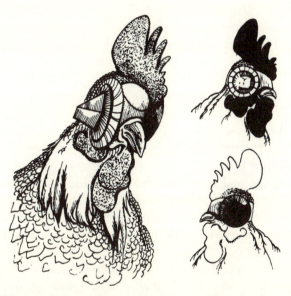

Figure 15 The results of experiments with disturbances of vision produced by prisms are different according to the species studied. While adaptation occurs quickly in most human subjects and in monkeys, the same is not true in less evolved species such as birds.

Around 1920, Pfister noted that chickens equipped with a prism which inverted the visual field of one eye (the other being covered) were incapable of precise pecking. No adaptation was noted. One of the possible explanations for this negative result was that the chicken was lacking in the capacity to adapt to this sort of sensorimotor conflict. Another, more reasonable, explanation was that, being unable to see the tip of its beak, the animal could not sense the separation between its peckings and the grains toward which they were directed; thus, no corrective mechanism could be brought into play.

and that our knowledge of the outside world came from it. For him, only the tactile experiencing of objects could furnish vision with the indices needed to appreciate distance and localization of objects in space. Berkeley's point of view was, however, purely theoretical[4] and left unanswered the famous question Molineux posed in a letter to Locke: Could someone blind from birth, who had learned to distinguish a sphere from a cube by touch, distinguish them without touching them if he regained his sight?[5] Throughout the nineteenth century, the surgical removal of cataracts in subjects born blind (which in turn gave them access to the visual world) reactivated this debate.

It is therefore not surprising that Stratton tried his hand at solving this. His idea, in disrupting the "harmony of sight and touch," was to see to what extent sight was led not to "accept the evidence" and align itself with contradictory information furnished by touch, rather than the reverse. In order to do this, Stratton wore glasses which inverted the visual field, for several days without interruption; objects were thus seen upside down, and their position in the visual field seemed modified, the objects which were normally located above now below and *vice versa*. Therefore, an object which fell to the floor seemed to fly into the air. At the end of about three days, Stratton noted a certain degree of adaptation of his vision to these new conditions, without being able to give a firm answer to the question he had asked regarding the respective roles of vision and touch. At the very least, the vertigo and total incoordination of his movements, which he had experienced during the beginning of the experiment every time he had tried to grasp an object, had partially disappeared. This experiment in radical transformation of the visual world has been recently repeated by H. Dolezal, who inflicted "reversing prisms" on himself continuously for seventeen days, and published his observations in detail.[6] Dolezal noted the importance of the disorganization of visuomotor behavior brought about by the prisms: When he wanted to grasp an object placed close to the ground he had to look up while at the same time lowering himself and directing his arm downwards! A great number of reflexes, such as those which normally coordinate the movements of the eyes with those of the head and body, were found, as a result of the prisms, to work at cross-purposes. Adaptation, which occurred in a few days, at first required much concentration on the subject's part, but was finally complete. At the end of his experiment, Dolezal could ride a bicycle, swim, and shake hands appropriately. He could precisely and correctly throw a ball (but could not catch it if it were thrown to him). Even more surprisingly, the reflex movements of the eyes with respect to the head were completely reversed by the end of approximately two weeks of wearing the prisms.[7]

Held's experiments, while practiced in a less ecologic context than those of Stratton and Dolezal, nonetheless demonstrated elements essential to the understanding of how new coordination responsible for adaptation comes about. It is, of course, impossible to describe this long series of results which appeared over fifteen years, starting in 1959.[8] The principal findings one can take from them are represented by two propositions. The first is that returning information about the executed movement (re-afferent information) must be systematically correlated with the movement. A subject can adapt to major sensorimotor conflicts (as Dolezal's experiment showed so well) on the condition that the conflict was constant, or that the deformation of the relationship between "position seen" and "position sensed" of the object remained systematic. In any other case, no adaptation was manifest. It was the same if a delay was artificially introduced between movement and its re-afference: when the delay was greater than one-fifth of a second, the precision of the movement decayed, as if its control had become impossible. A simple example allows one to prove this: a subject wearing earphones into which his own voice is projected becomes rapidly incapable of speaking if, by an electronic alteration, the emission and reception of the voice are displaced by only a few milliseconds. Clearly our own ecologic milieu does not expose us to this type of situation. It is quite different, however, in man–machine relationships, where conditions affecting control and execution of movements must remain within defined limits, so that the "human operator" will be able to adapt to them.

The second finding derived from Held's experiments is that recoordination in the course of a sensorimotor conflict produced by wearing prisms is not produced unless the subject executes movements in an active manner. This condition, as we have seen, comes from the explanatory model we dicussed at the beginning of this chapter. If the subject wearing prisms remains "passive" during the exposition of the conflict, adaptation does not result. One can thus compare the rapidity of recoordination in two subjects wearing prisms: one walks around the laboratory on his own, while the other is taken around in a wheelchair. In order to make the amount of displacement equivalent between the subjects, Held arranged the experiment so that the active subject himself pushed the wheelchair of the passive subject. Under these conditions, only the first of the two subjects had a complete recoordination of his movements, while the second remained in the previous coordination, as if he had never worn prisms.

These experiments demonstrated the high degree of adaptability of the sensorimotor system, even at the level of relatively simple reflexes, which one would have thought *a priori* organized in a more rigid manner. This property, which seems to belong to neither the sensory system nor the motor system as such, seems to be a characteristic of the mysterious *junction* where motor representation is established. The fact that movement must be active for this adaptability to be manifest shows that the pure sensations received passively by the organism do not really take part in this process: they only intervene as a logical complement of the active movements which produced them, as a dependent variable.

Perception is constructed by action: it is, as Bergson said, a particular property of sensorimotor activity. In this sense, Held's hypothesis opposes Gibson's point by point; Gibson's theory held that all the elements of the action are contained in the structure of the sensory world. If, as we saw in the preceding chapter, Gibson attributed a certain importance to the fact that the subject was displaced with respect to the visual scene, it was only to introduce the effects of gradients, parallax, and variation of the relationship between distance and size of objects. There is therefore nothing in this theory which makes the active displacement of the subject somehow special, when compared to passive displacement. The fact of exploring the environment augments the chances of contact between the subject and the texture of the visual world; however, even if this exploration is active, in Gibson's concept the subject remains a *sensory* subject, whose receptors directly record the effects due to the changes in perspective. There is nothing similar in Held's model, which, to the contrary, implies a sequential, point-by-point dependence of the motor and sensory events resulting from the intention which produced them. The critical role of active movement in these adaptation and recoordination processes thus constitutes a weighty argument in favor of the existence of a motor representation, whatever form one gives it. The possibility of modifying, in a lasting fashion (as the presence of an "after-effect" testifies) the internal processes which preside over the execution of movement is really the proof that these processes are real: What does not exist cannot be modified. The neurobiologist finds himself in a similar situation to that of a virologist during the 1950s, who, before being able to see a virus directly, could nonetheless affirm its existence by the very fact of the immunologic modifications "it" produced.

Gibson's supporters attempted to explain the effects of adaptation by a process of "normalization." In their conception, the visual environment consisted of a certain number of special axes, imposed in some way by the structure of the physical world: the vertical, for example, would be one of these axes, made special by the omnipresence of the gravitational field. The mechanism of perception would always seek to make the cardinal dimensions of the observed scene line up with these axes. One could thus show that prolonged observation by an immobile subject of a familiar scene (the interior of a room) which was tilted a few degrees off the vertical (by an optical system) would lend to an apparent realignment of the room. The effect of normalization was so strong that if one presented the same scene to an adapted subject, but this time oriented it correctly with respect to the vertical, it would appear tilted in the other direction! This observation did not, in fact, necessarily indicate that perception could be modified in a purely passive fashion. Nothing prohibited one from thinking that in the course of showing the tilted scene, the subject had, in fact, remained active: he could move his eyes, modify the position of his head, and thus assemble a sufficient number of re-afferences in order to adapt himself. Moreover, it could be shown in similar situations that deliberately active subjects became adapted more quickly and more completely than more "passive" subjects.

Could the notions concerning adaptation and recoordination also be applied to the normal acquisition of sensorimotor coordination, which is progressively established during the first years of life? This hypothesis, which was only implicit in our treatise up until now, was also the object of a long series of experimental studies by Held and his colleagues. These authors used the kitten, a very immature animal at birth which only acquires its principal visuomotor reactions near the age of three to five weeks. If, during this period, the young animal is kept in complete darkness, it appears to be completely naive with respect to the visual world at the age when any normal kitten has become a veritable sensorimotor machine; although in no way blind, it cannot avoid obstacles while walking, cannot jump off a footstool, and does not react to the rapid approach of an object. This state indicates that the visual experience acquired in the course of the first month is essential for establishing the visuomotor reactions of the adult animal.

In order to understand this better, Held had the idea of exposing kittens raised in the dark to light, but only for several hours each day. The kittens were grouped in pairs: in each pair, one of the animals was designated the active kitten, while the other was the passive kitten. During the course of the brief daily exposure to light, the active animal was allowed to move around freely, except that each of its movements was transmitted via a sort of harness, to which was attached the passive kitten. At the age of four weeks, a series of tests revealed that only the active animal had a normal development of its visuomotor reactions, while the other animal, although submitted to the same quantity of visual stimulation, remained naive at this level.

Another experiment, even more spectacular, seemed to give the key to understanding this. In this experiment, kittens were raised under normal visual conditions. However, they wore large collars around their necks, which did not allow them to see their own bodies, in particular their paws, in the course of their movements. Once again, it was noted that at the age of four weeks these animals had significant deficits when their visuomotor reactions were tested.

Two conclusions can be drawn from this work. On the one hand, it demonstrates the important role active experience plays during the course of early development in the acquisition of sensorimotor coordination. On the other hand, it confirms the importance of a systematic correlation between each movement and the effect it produces in the course of active experience.

If we have spent a great deal of time on Held's models and the experiments which resulted from it, it is in large part because of their theoretical importance. This work gives rise to a crucial discussion on the role of motor activity in the acquisition and development of the individual's capabilities. Sensualist empiricism, which we have run across at several turns, implies a passive development of these capabilities by accumulation, the edification of knowledge of the outside world being effected by sensory experi-

Figure 16 The famous device of Held and Hein (1963). All the movements of the "active" kitten were transmitted to the "passive" kitten. The amount of visual stimulation is thus the same for the two animals. However, only the "active" kitten received visual stimulation which was in direct causal relation with its own spontaneous motor activity, allowing it to develop its visuomotor coordination properly. From R. Held, "Plasticity in sensory-motor systems." Copyright © 1965 by Scientific American, Inc. All rights reserved.

ence. Constructivism is very different, and assuredly less naive; according to this theory, the subject progressively constructs his representation of the real through his action on the environment. This hypothesis, as we will see, is located between empiricism and nativism, which views development as the progressive revelation of innate forms, wherein the environment certainly intervenes, but more in the form of epigenetic material than in terms of an interaction.

It would seem interesting to place these relatively abstract theories next to the different neurobiologic models of learning and development, necessarily conceived upon a more realistic basis since they must take the structure of the nervous system into account. A number of these models were inspired by the work of Ramon Cajal, who hypothesized in 1909 that the growth of neural terminations did not stop at birth but continued subsequently under the impulse of use. Cajal started from the general idea that "the voluntary expansions of new creation do not proceed by chance; they must be oriented according to dominant neural currents, or in the direction of intercellular association which is the object of the reiterated solicitations of the will."[9] This hypothesis was reformulated in 1949 by the Canadian D. O. Hebb, who postulated that the synapses assuring contact between neurons had to be modifiable, and that the repeated excitation of one neuron by another could produce a "growth process," which would result in increasing the contact surface, and thus the efficiency of transmission, between the two neural elements.[10] Numerous experiments were performed on the basis of this model. Newborn animals were subjected to sensory deprivation (usually visual), more or less prolonged. One hoped to prove *a contrario* the validity of the model, by showing that placing the synapses at rest (in this case, those of the visual system) would result in a reduction of their volume and efficiency. The results of these experiments were the object of contradictory interpretations: on the one hand, some tended to credit the notion of a complete malleability of the nervous system, whose performance was established by a simple impregnation by the milieu. Others, however, took into account the necessity of interaction between the innate order of the neural connections and the formative role (in the strict sense) of the milieu on the capacity of these connections to fulfill their function. Following experiments involving vision deprivation in the cat, D. Hubel and T. Wiesel were able to establish that the formation of neural pathways and the corresponding capacities obeyed an innate mechanism, which had to be subsequently confirmed or validated by the exercise of the function.[11] It is at this level that the real debate between constructivism and nativism is located.[12]

Those psychologists who study very young children conclude that, on the whole, the development of the principal functions— motor functions, logic operations, language, etc.—proceeds by *stages*. The problem posed is, in fact, interpreting these stages. For the nativist, it is a question of a sudden actualization, when a certain degree of maturity is attained, from a set of preformed notions or given possibilities present from the beginning. What is undoubtedly true for the structural stages responsible for the embryonic, and later postnatal, formation of the nervous system might also be true for the stages of perfomance acquisition. The nativists themselves, and above all N. Chomsky, think that language starts with an innate "fixed nucleus," which contains the set of logical operations needed for learning language, but that later the environment intervenes to complete the process. The problem of the nature of this fixed nucleus can certainly be posed: Should one consider it to be like a set of "working hypotheses," as K. Lorenz said, which the subject had at its disposal to incorporate and structure information from the sensory world. Or, on the other hand, should one consider it an "intrinsic competence" which specifies the grammar accessible to the human infant, as Chomsky claims? Whatever solution is reached, the difference in conception of the epigenetic role attributed to the environment could never be anything but a difference of degree, since, in all cases, this role would be limited to furnishing the elements necessary to fill the gaps in the genome. Nativism thus inevitably leads to the idea of a *repertoire*, to take up the term used by ethologists, that is, the idea of a predetermination of the performances effectively available to the individuals of a species, among the multitude of possible performances.

In the constructivist perspective, the stages of development are authentic "constructions," opening each time on new possibilities. What is more, in the form given them by J. Piaget, these constructions result in states of equilibrium between exogenous factors (coming from the environment) and endogenous factors (determined by the genome). Thus, construction could model the structure of the nervous system in the course of individual development, by a mechanism of "auto-regulation."[13] But what do these constructions consist of? For Piaget, they expressed the fundamental fact "that all knowledge is tied to an action, and that one can only know an object or an event by using them through assimilation into schemes of action." Piaget continued, "To know does not consist of copying the real, but acting on it and transforming it . . . in order to understand as a function of the systems of transformation to which these actions are linked."[14] During the course of a first stage of development (the sensorimotor period), the infant uses schemes of action in order to coordinate his own actions with the necessities of the outside world. Later, he uses them to acquire logic operations or mathematics, etc. Thus, the scheme of action (that which, in an action, can be generalized from one situation to another, or what is common to the many repetitions or applications of this action) constitutes a factor for assimilating the basic elements of the sensory world, and for constructing the real.

These ideas, to which Held's experiments surely gave more consistency than the long developments of Piaget, have nonetheless not found their way into a neurobiologic model yet. In this model, motor activity must be taken into account with its original, particular characteristics, that is, with what it has that is most voluntary and "subjective." It is only at this level that it can become the true manifestation of the self-organizing power of the acting subject. Reflex motor activity is only determined by the necessity of responding efficiently to a stimulus; it obeys an imperative of adaptation to the environment. The reflex response is thus completely predictable, as far as its appearance and its form are concerned. It is a one-way process which exhausts itself in its own execution and which, as a result, does not generate new information for the organism which produces it. Spontaneous motor activity, on the other hand, while directed toward the external environment, does not depend upon it for its onset. It is, by definition, the expression of a new phenomenon, not directly predictable and which, by itself, carries information.

The apparently random character of the appearance of spontaneous movements, which confers on them their novelty, should not be misinterpreted. One can certainly conceive of physical phenomena which arise from a mechanism that is random by nature (even if they represent a challenge to the laws of the propagation of energy), and as a result are capable of spontaneous variations.[15] This formulation, however, would lead spontaneous movement back to being simple fluctuations in an energy system. To qualify as random, the production of movements which are not otherwise precipitated brings us back to recognizing our ignorance of the laws which control this production. In a speech given in 1872, Dubois-Reymond concluded that there are absolute limits to the knowledge of nature: "Ignoramus, ignorabimus"; we would never know the link between matter and energy, or between consciousness and movement.[16]

If the status of spontaneous movement—and its psychologic counterpart, voluntary movement—is just as difficult to discern, it is because spontaneous movement does not seem to correspond to any precise *function*. What good to the organism which contains it is a phenomenon which produces itself? The question, however, is not without object: it is posed from the very fact of the individualization of a category of movements which owe nothing to reflexes. If not, why individualize it; why not follow the path traced by so many predecessors, which tended to group all movements in one category?

The answer to these questions depends upon the degree of generality of the relationship which links spontaneous activity with the generation of information, such as it has been established in the discussions encountered in this chapter. The construction of sensorimotor coordination, its maintenance, and its springing forth during new situations, which correspond to adaptive necessities, need information coming from interaction with the environment. Logic would have the acquisition and stabilization of performances, whatever they may be, including the least "useful," obey the same rules. At the extreme, in this hypothesis, voluntary movement would become the special, if not unique, means destined to feed information to a machine which never ceases needing it in order to construct itself, then function.

The whole sense of spontaneous movement is there, in the conceptualization which neurobiologic thought imposes: It is not the environment which solicits the nervous system, models it, or reveals it. On the contrary, it is the subject and his brain which question the environment, inhabit it little by little, and finally master it.

NOTES
INDEX

Notes

1. The Natural History of the Soul

1. R. Descartes, *Traité de l'homme*. The work was published in Latin in Leiden in 1664. The citations reproduced herein were taken from Descartes, *Oeuvres philosophiques*, ed. Ferdinand Alquié (Paris: Garnier, 1963), I, 378f.

2. *Traité de l'homme*, p. 445.

3. The term "animal spirits" was progressively abandoned during the eighteenth century. From von Haller on, the term *vis nervosa* was used more and more. Later, the more modern terms "nervous principle" and "nervous influx" would appear. As for the nature of this force, before its electrical origin was definitively established, a long debate took place over Newton's ether theory versus Willis's fire theory. See G. Canguilhem, *La formation du concept de réflexe aux XVIIᵉ et XVIIIᵉ siècles* (Paris: Vrin, 1955; 1977), chap. 4.

4. Regarding the history of the electrical origin of the nervous influx, see M. A. B. Brazier, "The historical development of neurophysiology," in *Handbook of physiology*, sect. 1, chap. 1, pp. 1–58. As for the nervous fluid, this idea was taken up again much later. It is now known that a constant flux of protein occurs along the axon of the nerve cell. This flux, which is quite slow, has a trophic role and does not participate in the genesis of the nervous influx. Certain authors in the eighteenth century, R. Whytt in particular, had an intuition that the nervous fluid must circulate very slowly, given the extreme smallness of the nerves, and could not therefore explain a rapid phenomenon such as muscular contraction.

5. R. Descartes, *Discours de la méthode*, pt. 5 in *Oeuvres philosophiques*, I, 613f. For an analysis of this aspect of Descartes's thought and its role in the evolution of ideas in the eighteenth century, see E. Cassierer, *Die Philosophie der Aufklärung* (Tübingen: J. C. B. Mohr, 1932), esp. chap. 3.

6. *Discours de la méthode*, pp. 628–629.

7. The iatromechanicists thought they could reduce all physiologic and pathologic phenomena to mechanical actions. Certain of them attempted to construct mechanical models which could explain phenomena such as digestion.

8. J. O. de La Mettrie, *L'homme-machine* (Paris: J.-J. Pauvert, 1966), pp. 124–125.

9. Ibid., p. 128.

10. Ibid., pp. 128–129.

11. Ibid., p. 128.

12. D. Diderot, "Locke," *Encyclopédie*, IX, cited in R. Desné, *Les matérialistes français de 1750 à 1800* (Paris: Buchet-Chastel, 1965), p. 179.

13. Baron d'Holbach, *Le système de la nature*. The passage cited is reproduced in Desné, *Les matérialistes*, p. 95. Regarding d'Holbach, see P. Naville, *D'Holbach et la philosophie scientifique au XVIIIᵉ siecle* (Paris: Gallimard, 1967).

14. P. J. G. Cabanis, *Rapports du physique et du moral*, II: *Histoire physiologique des sensations*. Reproduced in part in Desné, *Les matérialistes*, pp. 195–196. In this report, published in 1802, Cabanis attempted to resolve the question: "Are the guillotined still conscious after decapitation?"

15. Ibid., p. 196.

16. Regarding the historical importance of F. J. Gall in the discovery of cerebral localization, see G. Lanteri-Laura, *Histoire de la phrénologie* (Paris: PUF, 1970). Also see H. Hécaen and G. Lanteri-Laura, *Evolution des connaissances et des doctrines sur les localizations cérébrales* (Paris: Desclée de Brouwer, 1977).

17. X. Bichat, *Recherches physiologiques sur la vie et la mort* (1800; Verriers: Gerard, 1973), p. 64.

18. Ibid., p. 12.

19. Ibid., p. 13.

20. Ibid., p. 72.

21. P. Flourens, *Recherches expérimentales sur les propriétés et les fonctions du système nerveux dans les animaux vertébrés* (Paris: Crevot, 1824; 2nd expanded ed., Paris: Baillière, 1842). The ideas of Flourens on the problem of materialism are more evident in his work *Histoire des travaux et des idées de Buffon*, 2nd ed. (Paris: Hachette, 1850). According to Flourens, Buffon's thought had numerous similarities with Descartes's, in particular when the two men dissociated the thinking soul from the body; it was in this, he said, that the philosophy of Descartes was "the great philosophy."

Why, then, did Descartes find it necessary to distance himself from his own orthodoxy when he invented animal spirits? Was this a language barrier? "Let us forget the small mechanism of the *animal spirits*, imagined by Descartes, let us forget all the small explanations of that which cannot be explained, the action of the brain, the ultimate action of the organ" (p. 117). As for Buffon, "he was scarcely any wiser. For *animal spirits*, he substituted the organic movements by which he explained everything. Let us leave these organic movements and let us get to the heart of the ideas" (p. 119).

22. J. B. Bouillaud, "Recherches cliniques propres à démontrer que la perte de la parole correspond à la lésion des lobules antérieurs du cerveau, et à confirmer l'opinion de M. Gall sur le siège de l'organe du langage articulé," *Arch. Gen. Med.* 8 (1825): 25–45. Reproduced in H. Hécaen and J. Dubois, *La naissance de la neuropsychologie du langage (1825–1865)* (Paris: Flammarion, 1969), p. 15.

2. Movement as Argument and Mechanism

1. G. L. Leclerc de Buffon, *Histoire naturelle de l'homme* (1749; Paris: Maspero, 1971), p. 44.

2. Regarding Vieussens, see J. Soury, *Le système nerveux central, structures et fonctions: histoire critique des théories et des doctrines* (Paris: Carvé et Naud, 1899), I, 446f.

3. Willis was the first to have affirmed the importance of the cerebral cortex, and to have made of its enfolding an attribute of the most evolved species, in particular man. However, he thought that the anatomic organization of the convolutions obeyed no precise rule, and had to remain variable from one individual to another in order to explain individual choice and changes in the exercise of animal functions.

4. Regarding Prochaska, see Canguilhem, *La formation*, chap. 4, "Unzer et Prochaska," and Soury, *Le système*, I, 465f.

5. J. J. C. Legallois, *Expériences sur le principe de la vie, notamment sur celui du coeur, et sur le siège de ce principe* (Paris, 1812). Released again in 1824 in *Oeuvres*, I, 64–65.

6. The use of the term "voluntary movement" to designate the whole of the movements of the extremities, as opposed to the "involuntary" motor activity of the organs, is obviously an abuse of language which served only to maintain confusion. One can also see in this a sign of the progressive occultation of the true nature of voluntary movement, which left the arena of physiology without ever really having been there.

7. *Expériences sur le principe de la vie*, pp. 135–136.

8. A. Vulpian, *Leçons sur la physiologie générale et comparée du système nerveux* (Paris: Baillière, 1866).

9. Charles Bell used rather dubious methods in this affair, going so far as to change his text of 1811, when it was reissued, in order to prove that he thought of the polarization of the posterior roots first. Flint, in 1868, published the two texts side by side and revealed the manipulation. A. Flint, "Historical considerations regarding the properties of spinal nerve roots," *J. Anat. Physiol.* 5 (1868): 521–538, 579–592. Even changed, Bell's work was not convincing. After having recognized this, Magendie stated that "Mr. Bell was close to discovering the functions of the spinal roots, led by his ingenious ideas regarding the nervous system; nonetheless, the fact that the anterior roots were designed to control movement while the posterior roots were more particularly associated with sensation, seems to have escaped him" (quoted by Flint, p. 588).

10. For a rehabilitation of Hall, consult F. Fearing, *Reflex action: a study in the history of physiological psychology* (Baltimore: Williams and Wilkins, 1930; Cambridge: MIT Press, 1970), chap. 9, "Marshall Hall."

11. *Leçons*, p. 394.

12. For Pflüger, the relationship between the individual and his environment could not be regulated by a pre-established harmony. An autoregulation, based on the needs of the individual, must intervene. Pflüger made explicit reference to feedback regulation mechanisms, which were to become one of the bases of cybernetics. See V. Henn, "The history of cybernetics in the 19th century," in *Pattern recognition in biological and technical systems* (Berlin: Springer Verlag, 1971).

13. C. Bernard, *Leçons sur les phénomènes de la vie communs aux animaux et aux végétaux* (Paris: Ballière, 1878; Paris: Vrin, 1966), pp. 113–114.

14. Vulpian, *Leçons*, p. 414.

15. See Fearing, *Reflex action*, chap. 11, "The Pflüger-Lotze controversy."

16. Vulpian, *Leçons*, p. 419.

3. Hierarchy and Integration of Movements

1. *Reflex action*, p. 253. On the history of neurology, see the excellent book by J. D. Spillane, *The doctrine of the nerves* (Oxford: Oxford University Press, 1981).

2. M. Merleau-Ponty, *La structure du comportement* (Paris: PUF, 1942; 3rd ed., 1953), p. 7. To see some sort of intention in the order, prospective

aspect of the execution of a reflex was, for Merleau-Ponty, pure anthropomorphism. "If behavior seems intentional, it is because it is governed by certain pre-established nervous pathways, such that I obtain the goal intended" (p. 7).

3. Ibid., p. 19.

4. Ibid., p. 22.

5. Ibid., p. 33.

6. C. Darwin's *The origin of species* was first published in 1859.

7. C. Darwin, *The descent of man* (London: J. Murray, 1871). Darwin showed a certain reserve toward Spencer, by leaving open the possibility of the appearance of a particular characteristic, the voluntary movement, which owed nothing to the reflex. "Spencer maintains that the first glimmers of intelligence are developed by multiplication and coordination of reflex actions . . . I am nonetheless quite reluctant to deny that instinctive reactions might be able to lose their fixed, natural character and be replaced by others, accomplished by the free will."

8. See H. Ey, *Des idées de Jackson à un modèle organo-dynamique en psychiatrie* (Toulouse: Privat, 1975). This work is in large part a reissue of a text of 1938, H. Ey and J. Rouart, *Essai d'application des principes de Jackson à une conception dynamique de la neuro-psychiatrie* (Paris: Douin).

9. *Des idées de Jackson*, p. 55.

10. The existence of peripheral inhibitory nerves was known after Weber demonstrated cardiac slowing by vagal stimulation, and after Pflüger demonstrated inhibition of intestinal peristalsis by stimulation of the splanchnic nerve.

11. I. M. Setchenov, *Reflexes of the brain* (St. Petersburg, 1863; Eng. trans., Cambridge: MIT Press, 1965), p. 20. For a consideration of Setchenov's work, see A. Masucco Costa, *Psychologie soviétique* (Paris: Payot, 1977), chap. 2, "Ivan Setchenov."

12. Ibid., p. 42.

13. The autobiographical notes were printed along with *Reflexes of the brain*, pp. 111–112. Setchenov's ideas were considered antireligious and dangerously materialist for society. Having died in 1905, he never knew the triumph his teachings would experience under the new Russian regime. His name was added to the title of the principal Russian journal of physiology.

14. *Reflexes of the brain*, p. 43.

15. Ibid., p. 78.

16. Ibid., p. 106.

17. Ibid., p. 102.

18. I. P. Pavlov, *Experimental animal psychology and psychopathology* (Madrid: Internal Congress of Medicine, 1903). This text is reproduced in I. P. Pavlov, *Les réflexes conditionnés, étude objective de l'activité nerveuse supérieure des animaux* (Paris: PUF, 1927; 1977), p. 30.

19. *Les réflexes conditionnés*, p. 250.

20. Ibid., p. 224.

21. Ibid., p. 243.

22. See below, Chapter 7.

23. *Les réflexes conditionnés*, p. 156.

24. Habituation is the progressive diminution of a reflex response when the stimulus is repeated many times. Pavlov observed that a new stimulus (a noise) precipitated an orientation (movement of the head and the eyes, increase in the heart rate). After several repetitions, the response disappears, but then reappears as soon as the stimulus is changed, as by increasing the intensity, for example. In the same fashion, the withdrawal reaction seen in a snail as a result of cutaneous stimulation disappears if the stimulus is repeated. This phenomenon of habituation follows the same way as in the higher vertebrates.

25. Cited by M. A. B. Brazier, "Historical development." Also see M. A. B. Brazier, "La neurobiologie, du vitalisme au materialisme," *La Recherche* 8 (1977): 965–971.

26. E. G. T. Liddell, *The discovery of reflexes* (Oxford: Oxford University Press, 1960).

27. C. S. Sherrington, *The integrative action of the nervous system* (New York: Charles Scribner's Sons, 1906), p. 8.

28. *Expériences sur le principe de la vie.*

29. C. S. Sherrington, *The brain and its mechanisms* (Cambridge: Cambridge University Press, 1933). Cited by M. A. B. Brazier, "Historical development."

30. K. Goldstein, *Der Aufbau des Organismus* (The Hague: M. Nijhoff, 1934). French trans., *La structure de l'organisme* (Paris: Gallimard, 1951), p. 68.

31. *La structure de l'organisme*, p. 76.

32. Ibid., p. 76.

33. See the work of M. Jeannerod and H. Hécaen, *Adaptation et restauration des fonctions nerveuses* (Villeurbanne: Simep, 1979), chap. 3, "Compensation des modifications périphériques du système moteur."

34. *La structure de l'organisme*, p. 191.

35. The work of K. Popper and J. C. Eccles, *The self and its brain* (Berlin: Springer Verlag, 1977), is a typical example of this "new" dualism.

4. The Motor Cortex

1. Cited by Soury, *Le système*, p. 473.

2. Ibid., p. 579.

3. "Recherches cliniques propres à démontrer que la perte de la parole," p. 30.

4. Ibid., p. 30.

5. Sixty years later, D. Ferrier would make the same criticism with respect to Flourens, "whose experiments in this area [the physiology of movement] were the origin of so many erroneous conclusions." It was true, according to Ferrier, that "frog and pigeon physiology has too often been the bane of clinical medicine, and has contributed to the discrediting of a method of research which, employed wisely, is . . . the cornerstone of exact biologic and therapeutic research."

6. The question of knowing whether each point of the motor cortex controls the contraction of a muscle or the execution of a movement may be dated from this era. Now, it is thought that the former hypothesis is the correct one, and that only a very minimal degree of coordination or synergy between the different segments involved in a complex movement can be realized by the motor cortex proper. Moreover, the motor cortex is divided into several zones: The motor cortex in the strict sense (area 4) controls the force of muscular contraction, while the zones surrounding it (area 6 and supplementary motor area) are responsible for more complex operations, such as initiating multisegmentary movements.

7. Ferrier was aware of the possible effects of current diffusion beyond the stimulated point, any such diffusion diminishing the precision of the results. In order to measure this, he used the physiologic rheoscope of Dupuy: A frog nerve-muscle preparation was placed across the posterior pole of the cortex; if the muscle contracted at the instant of current application to any other point on the cortex, one could deduce that the current had diffused.

8. D. Ferrier, *The localization of cerebral disease* (London: Smith, Elder, 1878).

9. Ibid., p. 34.

10. Regarding F. Goltz, see J. Soury, *Les fonctions du cerveau doctrines de l'école de Strasbourg, doctrines de l'école italienne* (Paris: Lecrosnier et Babe, 1891), p. 21.

11. Cited by Soury, ibid., p. 24.

12. Ibid., p. 28.

13. *The localization of cerebral disease*, pp. 35–36.

14. Goltz, describing a dog with a left hemisphere cortical lesion,

wrote: "If, in talking to him and touching his right paw, I ask the dog to give me his paw, I can readily ascertain by the expression on his face that he understands my order, and if, at last, as if in desperation, he offers me his left paw, I can see that the animal made every effort to satisfy me, but could not do what he was told. Between the organ of the will and the nerves which execute this will, an insurmountable resistance has been raised." What better definition could one wish for voluntary movement?

15. Cited by Soury, *Le système*, I, 615.

16. Maurice Schiff thought that voluntary movements were, in fact, simple reflexes and that the paralysis of a cortical lesion was a result of the loss of tactile sensation. However, the discussion of his theory included such contradictions, in particular as regards the anatomy, that it did not stand up to the attacks of its detractors.

17. Regarding C. Bastian, see E. G. Jones, "The development of the 'muscular sense' concept during the nineteenth century and the work of H. Charlton Bastian," *J. Hist. Med. All. Sci.* 27 (1972): 298–311, esp. p. 302.

18. Cited by Soury, *Le système*, II, 1100.

19. François Franck, *Leçons sur les fonctions motrices du cerveau (réactions volontaires et organiques) et sur l'épilepsie cérébrale* (Paris: Doin, 1887), pp. 368–369. Regarding the notion of motor images, see chap. 5.

20. F. W. Mott and C. S. Sherrington, "Experiments upon the influence of sensory nerves upon movement and nutrition of the limbs," *Proc. Roy. Soc.* 57 (1895): 481–488, esp. p. 483.

21. The effects observed by Mott and Sherrington, at first confirmed during the 1950s, have more recently been called into question. Similar experiments performed from 1968 on have shown that complete loss of sensation in an extremity does not prohibit the production of voluntary movements. Even more impressive are experiments in which the posterior roots are sectioned at birth, or *in utero*, in a monkey. These animals subsequently develop a motor function that is close to normal, only delayed by a few months with respect to a normal animal. The current data demonstrate that the loss of sensation in an extremity *does not inhibit* the production of voluntary movements by this extremity, but the execution of these movements is very defective. See M. Jeannerod and H. Hécaen, *Adaptation*, chap. 4, "Effets de la désafférentation somesthésique sur le contrôle des mouvements." See also M. Jeannerod, F. Michel, and C. Prablane, "The control of hand movements in a case of hemianesthesia following a parietal lesion," *Brain* 107 (1984): 899–920.

22. Soury, *Les fonctions du cerveau*, p. 237.

23. Ibid., p. 379.

24. Ibid., p. 377.
25. Ibid., p. 377.

5. The Motor Idea?

1. There are numerous neural pathways other than the pyramidal tract which arise from the motor cortex. These pathways, which are directed toward the cerebellum or toward the subcortical nuclei known as "the central grey nuclei" or "basal ganglia," constitute the first part of a circuit which returns to the motor cortex. Thus, information regarding a movement about to be executed can be used to control its own execution. The cortical-subcortical pathways represent an important contingent of fibers (several tens of millions). By comparison, the pyramidal tract, to which we have given such an important place in this discussion, is, numerically speaking, a small bundle (in man it contains only 1 million fibers). Among these, only 2 percent represent axons from the famous giant cells of Betz.

2. The model of an algebraic sum of activating (+) and inhibiting (−) effects is perfectly valid at the level of the synapses in a nerve cell network. This model, however, owes nothing to Pavlov and is a result of work on the activity of single nerve cell recordings (in particular, the works of Eccles during the period 1940–1950). To transpose it to another level of activity in the nervous system, in particular to a psychologic level, would lead back to the worst sort of reductionism.

3. *La structure du comportement*, p. 57.

4. *La structure de l'organisme*, p. 26.

5. Ibid., p. 27.

6. Ibid., p. 28.

7. Ibid., p. 28.

8. K. Goldstein is located chronologically in the heart of an antilocalizationist period. Karl Lashley, following a series of experiments regarding cortical ablation in the rat, thought that the effects of a lesion depended less on its localization than on its extent. Looking at his results, he came to the conclusion that the different cortical zones were equipotent with respect to "superior" nervous function, and that only the quantity of tissue involved mattered. He explained the phenomena of functional recovery following a lesion in this way. (See M. Merleau-Ponty, *La structure du comportement*, pp. 66f. See also H. Hécaen and G. Lanteri-Laura, *Evolution des connaissances*.)

9. E. Cassirer, "Etude sur la pathologie de la conscience symbolique," *J. Psychol.* (1929): 289–336, esp. p. 300.

10. H. Jackson, "On the nature of the duality of the brain," *Medical Press and Circular* 1 (1874): 19. Rpt. in *Brain* 38 (1915): 80–103, esp. p. 90.

11. Jackson, ibid., p. 91.

12. Quoted by Cassirer, 1929.

13. Jackson, "On the nature of the duality," p. 81.

14. C. Wernicke, *Der Aphasische Symptomencomplex* (Breslau, 1874). French trans. by E. Carstanjen, *Le complexe symptomatique de l'aphasie* (Paris: Ecole pratique des Hautes Etudes, 1974), p. 2. This concept was essentially due, at least in its anatomic-functional correlation, to Th. Meynert, who taught in Vienna. Before becoming professor at Breslau, Wernicke was Meynert's student for two years. Meynert was also a strong influence on Freud.

15. *Le complexe symptomatique de l'aphasie*, p. 3. It is important to note that this theory of learning is still at the core of most modern theories. Wernicke also noted that the nerve cells had the capacity to be modified in a lasting fashion after an excitation, and that this modification was molecular in origin. See M. Jeannerod and H. Hécaen, *Adaptation*, chap. 7, "Bases neuronales de l'apprentissage," as well as the last chapter of this text.

16. *Le complexe symptomatique de l'aphasie*, p. 9.

17. Ibid., p. 10.

18. Ibid., p. 10.

19. Where Bastian postulated a muscular sense in order to explain the formation of a motor image, Wernicke postulated a "sensation of innervation" associated with the execution of a reflex movement. This sensation of innervation was propagated from the reflex center to the motor cortex by an association pathway. Regarding the difference between the muscular sense and the sensation of innervation, and the implications of these two concepts, see Chapter 6.

20. "Etude sur la pathologie de la conscience symbolique," p. 303.

21. Cited by J. L. Signoret and P. North, *Les apraxies gestuelles* (Paris: Masson, 1979), p. 26.

22. J. Dejerine, *Sémiologie des affections du système nerveux* (Paris: Masson, 1900; 1914; 1977), pp. 42–43.

23. Ibid., p. 43.

24. N. Geschwind, "Disconnexion syndromes in animals and man," *Brain* 88 (1965): 585–644, esp. p. 638.

25. Sperry's experiments were carried out in human subjects who had undergone a section of the corpus callosum (in an attempt to treat recalcitrant epilepsy). By conceiving of very simple tests, Sperry was able

to produce evidence for deficits caused by sectioning the corpus callosum, which were not evident on a routine exam. It was sufficient to be able to present a stimulus to one of the hemispheres, and record the response of either that hemisphere or the other, the hypothesis being that any transfer of information from one hemisphere to the other was, of necessity, made through the commissures, and thus that no transfer was possible when they were sectioned. A spectacular example of the results obtained using this methodology was the perception of visual "chimeras" by callosectomized subjects. These chimeras are images composed of two halves of different objects, constructed in such a fashion that one half falls into each visual hemifield. In this case, each hemisphere "recognizes" the object whose half is represented in the corresponding visual hemifield, and "ignores" the other half. The subject gave a response depending upon which pathway was used. The verbal response was the name of the object whose semi-image was represented in the right hemifield, since in the majority of subjects the left hemisphere contained the zone specialized for oral language. On the other hand, the right hemisphere could express itself by having the left hand choose from among a group of objects the one represented in the left hemifield. Through observations of this type, it was possible to demonstrate that the same sensory information gave rise to (at least) two modes of mental operations which the disconnection of the two hemispheres artificially isolated. The left hemisphere was responsible for an analytic treatment of the information, thanks to a sequential decoding of the different properties of the object, and expressed itself through language. The right hemisphere, on the other hand, was responsible for a more synthetic appreciation of the nature of the stimulus, by accenting the spatial aspects rather than the temporal aspects of the information represented.

26. The most recent work tends to show that the convergence of several sensory modalities on a single cortical area is much more limited than had been thought. This notion is still retained for a limited region of the parietal cortex (area 7); in monkeys, neurons receiving visual and somesthetic information, as well as information concerning the execution of ocular movements and movements of the arm directed toward extrapersonal space, are seen here.

6. The Physiology of the Will

1. Maine de Biran, *Mémoire sur la décomposition de la pensée* (1805). Cited from the Tisserand edition of *The works of Maine de Biran* (Paris: Alcan, 1930), III, 197.

2. Ibid., pp. 197–198.

3. Maine de Biran, *Influence de l'habitude sur la faculté de penser* (1803), ed. Tisserand, *Works*, III, 16. Regarding the philosophy of Maine de Biran and its ontologic aspects, see G. Le Roy, *L'expérience de l'effort et de la grâce chez Maine de Biran* (Paris: Boivin, 1937), and M. Henry, *Philosophie et phénoménologie du corps* (Paris: PUF, 1965), chap. 2, "Le corps subjectif."

4. Cited by G. Le Roy, *L'expérience*, pp. 196–197.

5. A. Schopenhauer, *Ueber den Willen in der Natur* (Frankfurt, 1836). French trans., *De la volonté dans la nature* (Paris: PUF, 1969), p. 79.

6. Cited by Th. Ribot, *La philosophie de Schopenhauer*, 7th ed. (Paris: Alcan, 1898), p. 65. Ribot had, moreover, developed an application of the theses of Schopenhauer to psychopathology in his work *Les maladies de la volonté* (9th ed., Paris: Alcan, 1894). Ribot only retained Schopenhauer's "general" conception of the will. According to him, the voluntary act is "the result of the entire neural organization," and not "the simple transformation of a state of consciousness into movements. Rather . . . it supposes the participation of this whole group of conscious or subconscious, states which constitute the self at a given moment." In other words, "from the lowest reflex to the highest will, the transition is very gradual, and . . . it is impossible to say exactly in what moment the will proper begins, that is, the personal reaction" (*Les maladies de la volonté*, pp. 32–33). It is not surprising that Ribot included, among the maladies of the will, a large part of psychopathology, from "weakenings of the will" created by depressive states to "excesses of impulsiveness" in obsessive neuroses or hysteria.

7. *De la volonté dans la nature*, p. 79.

8. Ibid., p. 80.

9. Regarding notions of the mental representation and the movement program, see Chapter 8.

10. During the same era, Ernst Brucke, who was to have a great influence on Freud, considered physiology as an extension of physics. He thought that organisms were particular objects, but could nonetheless be described as isolated systems wherein energy remained constant. Regarding this subject, see P. L. Assoun, *Introduction à l'épistémologie freudienne* (Paris: Payot, 1981), pp. 99ff.

11. Bain's ambiguity (who undoubtedly did not know Maine de Biran) reflects the discussion of the cause of voluntary movements by the English philosopher Hume. According to Hume, the will could only be considered objectively as the "cause" of movements to the extent that the effect (the movement) could normally be predicted from "the energy of its cause." "It is for this reason that we are condemned for eternity to never know

the efficacious means by which this most extraordinary operation is effected." See also M. Henry, *Philosophie*, regarding this.

12. Cited by P. L. Assoun, *Introduction*, p. 146.

13. Maine de Biran, *De l'aperception immédiate* (1807; Paris: Vrin, 1963), p. 54.

14. Ibid., p. 55.

15. A. Brodal, "Self-observations and neuro-anatomical considerations after a stroke," *Brain* 96 (1973): 675–694.

16. The action of curare on the neuromuscular junction was discovered almost simultaneously by Bernard and Kolliker, around 1850.

17. One could also cite the impressions of movement of the "phantom limb" in patients who have undergone amputation. In an editorial appearing in England in 1875 (unsigned, but attributed to Jackson), there was question as to whether certain of these subjects could have a very precise sensation of the movements or position of the hand at the end of the phantom arm. "My hand is now open, it is closed . . . I touch the thumb with the little finger . . . It is now in writing position." "In other words," concluded the editorialist, "the volition to move certain parts [of the body] is accompanied by a mental condition which represents the quantity of the movement, its force, and idea of a change in position of these parts to the consciousness." "Psychology and the nervous system," *British Medical Journal*, 9 October 1875, p. 462.

18. G. H. Lewes, "Motor-feelings and the muscular sense," *Brain* 1 (1878): 14–28, esp. p. 15.

19. Cited in H. E. Ross and K. Bischof, "Wundt's views on sensations of innervation: a reevaluation," *Perception* 10 (1981): 319–329.

20. The duality of the sensation accompanying a movement was remarkably discussed by Lewes. "I move my fingers by a voluntary effort; you then move my fingers in the same directions, I neither opposing nor resisting the movement. In both cases the effect is the same; but in the first case there is present what is absent from the second, namely, the element of effort. Now suppose that I am suffering from cutaneous anaesthesia, I shall then feel the effort, but not the effect; so that unless I am looking at my fingers, I shall be unable to say whether I have moved them or not. Herein we see disease analyzing the two elements of a muscular feeling, active and passive, central impulse and peripheral reaction." "Motorfeelings," p. 28.

21. Ibid., p. 24.

22. Ibid., p. 24.

23. Descartes had imagined this experiment and had illustrated it in a figure in his *Traité de l'homme* (p. 430).

24. H. von Helmholtz, *Handbuch der physiologischen Optik* (Leipzig: Vos, 1866). Eng. trans., *Helmholtz's treatise on physiological optics* (New York: Dover, 1962), III, 245–246.

25. W. James, *The principles of psychology* (1890; New York: Dover, 1950), vol. II, chap. 26, "Will," p. 508.

26. R. W. Sperry, "Neural basis of the spontaneous optokinetic response produced by visual inversions," *J. Comp. Physiol. Psychol.* 43 (1950): 482–489.

27. Ibid., p. 488.

28. E. von Holst and H. Mittelstaedt, "Das Reafferenzprinzip. Wechelwirkung Zwischen Zentralnervensystem und Peripherie," *Naturwis* 37 (1950): 469–476. The texts of von Holst, written for the most part in German, were translated into English and published as *The behavioral physiology of animals and man: selected papers of E. von Holst* (Coral Gables: University of Miami Press, 1973), p. 140.

7. The Physiology of Spontaneity

1. S. Freud, "An outline for a scientific psychology," manuscript of 1895, reproduced in *La naissance de la psychanalyse* (Paris: PUF, 1956; 3rd ed., 1973), p. 317.

2. Ibid., p. 320.

3. Ibid., p. 321.

4. Ibid., p. 353.

5. In 1950 Loewenstein demonstrated that the neurons of the brainstem receiving labyrinthine afferents were the site of a permanent (tonic) activity, even when no stimulation was applied to the receptor. Later, it was noted that this tonic activity disappeared after sectioning the nerve. This experiment proved the peripheral origin of spontaneous activity in "central vestibular" neurons. Regarding the history of the idea of reticular tone, see M. Jeannerod, "La découverte de la fonction vestibulaire," *Acta oto-rhino-laryngol. belg.* 35 (1981): 975–1006.

6. It is the same for spontaneous motor activity. Certain butterflies, who fly around in a lighted room, will immediately fall to the ground if the lights are turned out. There seem to be numerous other examples of this phenomenon of "photokinesis." Regarding the tonic function of sensory organs, see D. L. Meyer and T. H. Bullock, "The hypothesis of sense-organ-dependent tonus mechanisms, history of a concept," in "Tonic functions of sensory systems," *Ann. N.Y. Acad. Sci.* 290 (1977): 3–17.

7. Flourens, *Recherches expérimentales*, p. 32.

8. See N. Kleitman, *Sleep and wakefulness* (Chicago: University of Chicago Press, 1939). On the historic bases of the theory of sleep as deafferentation, see G. Moruzzi, "The historical development of the deafferentation hypothesis of sleep," *Proc. Amer. Philos. Soc.* 108 (1964): 19–28.

9. F. Bremer, " 'Cerveau isolé' et physiologie du sommeil," *C.R. Soc. Biol., Paris,* 118 (1935): 1235–1241.

10. See G. Moruzzi, "Historical development."

11. The absence of wakefulness, produced by a large lesion of the brainstem or a drug intoxication, is known as coma. Coma is distinguished from sleep by a multitude of criteria, including EEG criteria. M. Jouvet is in large part responsible for the fact that the mechanisms of sleep are now known. His principal discoveries go back to 1959, when he was able to demonstrate that a particular stage of sleep ("paradoxical sleep") was under the control of well-defined nuclei in the lower brainstem. In order to explain the spontaneous appearance of this state in the course of the sleep cycle, Jouvet made recourse to a "humoral" theory which made paradoxical sleep a response to modifications in the internal environment. It seems that the phases of "paradoxical sleep" correspond, in man, to those moments when one is having the subjective experience of dreaming.

12. A selection of the principal works of R. Hess recently appeared in English translation: *Biological order and brain organization: selected works of R. Hess,* ed. K. Akert (Berlin: Springer Verlag, 1981).

13. H. R. Hess, "Der Schlaf," *Klin. Wochenschr.* 12 (1933): 129–134. Cited from *Biological order*, p. 125.

14. H. R. Hess, "Biomotorik als Organisations problem I, und II," *Naturwis* 30 (1942): 441–448, 537–541. Cited from *Biological order*, p. 264.

15. See R. W. Doty, "Electrical stimulation of the brain in behavioral context," *Ann. Rev. Psychol.* 20 (1969): 289–320.

16. J. M. Delgado, *Physical control of the mind: toward a psychocivilized society* (New York: Harper, 1971), pp. 184–185. The title of this book is terrifying. The technique of electrical stimulation of the brain in animals permitted the observation of zones where excitation interrupted a movement in midstream, or inhibited the manifestation of a behavior. Delgado and his collaborators applied this technique in man. They describe the case of a subject who had "crises of antisocial conduct during which [he] attacked members of his own family." His episodes "were considerably diminished by repeated stimulations of the amygdaloid nucleus (p. 176). Other techniques, arising from the same idea of behavioral control, such as psychosurgery, use the same justifications and raise the same questions.

17. Ibid., p. 62.

18. Ibid., p. 62.

19. K. Lorenz, *Le tout et la partie dans la société animale et humaine: un débat méthodologique* (1950). Reproduced in *Essais sur le comportement animal et humain: les leçons de l'évolution de la théorie du comportement* (Paris: Le Seuil, 1970), pp. 321–322.

20. Ibid., pp. 328–329.

21. Ibid., p. 329.

22. The theory has an undeniable heuristic value. Bernstein was able to show that all the trajectories of the arm, except for the execution of complex repetitive movements, could be reconstructed from three fundamental, sinusoidal curves and their first three harmonics. The ideas of E. von Holst are discussed in "Van Wesen der Ordnung im Zentralnervensystem," *Naturwis* 25 (1937): 625–631, 641–647. English trans. in *Behavioural physiology of animals and man*.

23. As always in matters relating to the determination of the origin of a biologic phenomenon, the explanation is a bit more complex. Today, it is believed that the rhythmic activity of cells is due to an oscillation occurring spontaneously in the resting potential of the membrane.

24. See M. A. B. Brazier, "Historical development."

25. The technical difficulties experienced in obtaining this type of recording are of two sorts. It is first necessary to detect the potentials (which are very low voltage) from amid the "background noise" of the rest of the cerebral electrical activity. To do this, a summation is performed using a computer, the potentials corresponding to a large number of identical movements. As can be readily seen, there is a limit to this technique imposed by the repetition of the same voluntary movement by the subject. The second difficulty arises from the fact that since the movement is executed in a spontaneous manner, the exact onset of the electrical changes cannot be signaled to the computer. This difficulty can be surmounted by using the muscular discharge produced by the execution of the movement as a signal, and then resetting the time from this. This is possible thanks to the computer, which can store several seconds worth of electrical activity in its memory and then refer back to this.

H. Kornhuber, who perfected this technique in 1965, called the potentials which were linked to the preparation of a voluntary movement *Bereitschaftspotential*, which points to their relationship with the spontaneous nature of the action.

26. J. C. Eccles, *The humnan mystery*, Gifford Lectures, 1978 (Berlin: Springer Verlag, 1979), p. 255 in the French ed.

8. Representation, Plan, Program

1. H. Bergson, *Matière et mémoire: essai sur la relation du corps à l'esprit* (1896; 3rd ed., Paris: Alcan, 1903), p. 16.

2. Ibid., p. 34.

3. Ibid., p. 94.

4. Ibid., p. 251.

5. Moreover, Bergson made explicit reference to the vital nature of certain properties resulting from the cerebral activity when he approached the problem of memory. No more than in the matter of perception, it was not necessary to look for arguments in favor of an exact localization for the accumulation of memories in the cortical substance. In memory disturbances, "it is not this or that memory which is torn away from its localized site, but rather it is the faculty of recall which is more or less diminished *in its vitality*," as if the subject could not "connect his memories with the current situation." It was the role of the brain to assure the functioning of the mechanism of this contact, "and not to imprison the memories themselves in its cells" (ibid., p. 265).

6. Cited by S. Ullman, "Against direct perception," *Behav. Brain Sci.* 3 (1980): 373–415. Gibson's theory is discussed in *The senses considered as perceptual systems* (Boston: Houghton Mifflin, 1966).

7. Cited by Ullman, "Against direct perception."

8. Regarding behaviorism, see B. F. Skinner, *Beyond freedom and dignity* (New York: Knopf, 1971). The theory itself dates back to the work of J. B. Watson in the United States, around 1913. Watson, reacting against the psychology of introspection, stated that the only basis for psychology must be behavioral responses, and not the alleged internal states supposed to be at the origin of these behaviors.

9. See J. Requin, "Spécificité des ajustements préparatoires à l'exécution du programme moteur," in H. Hécaen and M. Jeannerod, *Du controle moteur à l'organisation du geste* (Paris: Masson, 1977).

10. U. Neisser, *Cognition and reality* (San Francisco: Freeman, 1976). "Cognition" in English and "cognitio" in Latin, translate as the action of learning in order to know, the intellectual process which allows one to acquire knowledge by perception or ideas.

11. This method of classification is quite similar to Jackson's, as was discussed in Chapter 3. Jackson instituted an approach currently known as "systems theory."

12. The problem of knowing whether complex systems such as computers could possess mental capacities was explicitly posed by A. Turing.

If a computer could convince experts in information science that it had such capacities, it would therefore have these capacities. Regarding the problems posed by artificial intelligence, see D. R. Hofstadter, *Gödel, Escher, Bach* (New York: Basic Books, 1979), and D. R. Hofstadter and D. C. Dennett, *The mind's I: fantasies and reflections on self and soul* (New York: Basic Books, 1982).

13. Regarding the relationship between the brain and the mental states, and the "functionalist" approach, see J. Fodor, *The language of thought* (Cambridge: Harvard University Press, 1974).

14. N. Bernstein, *The coordination and regulation of movements* (Oxford: Pergamon Press, 1966), p. 37.

15. Bernstein mentioned that the hypothesis of serialization resembles the associationist doctrine of the psychologists, while the hypothesis of parallel function conforms better to the notion of the will. While he affirmed that he did not wish to take part in them, he felt that "the attacks against the opinions of the pure associationists were well-founded" (ibid., p. 39).

16. This hypothesis leads us back to the idea that the representation of an action is that of the *goal to be obtained*, and not of the elementary constituents. "When I voluntarily start to walk," said Woodworth in 1906, "my intention is not that of alternately moving my legs in a certain manner; my will is directed toward reaching a certain place. I am unable to describe with any approach to accuracy what movements my arms or legs are to make; but I am able to state exactly what result I intend to accomplish." (Cited by G. A. Kimble and C. C. Perlemutter, "The problem of volition," *Psychol. Rev.* 77 (1970): 361–384, esp. p. 369. The discussion recalls a similar one regarding the representation of the muscles versus the movements in the motor cortex (see Chapter 4, n. 6).

17. Cited by V. Henn, *Pattern recognition*.

9. Action, the Force of Self-Organization

1. R. Held, "Exposure-history as a factor in maintaining stability of perception and coordination," *J. Nerv. Ment. Dis.* 132 (1961): 26–32.

2. The context itself authorizes such interpretations. Held, who cited Helmholtz, seemed to have nonetheless been quite influenced by the German school of the 1920s and 1930s. One cannot underestimate the role of H. L. Teuber, a disciple of K. Goldstein who worked in Germany and later in the United States, and who was an ardent propagandizer for the idea of corollary discharge. When Teuber became the director of a psychology laboratory in 1962, he assured himself of Held's cooperation.

3. H. von Helmholtz, *Optique physiologique*.

4. G. Berkeley, *An essay towards a new theory of vision* (1709), rpt. in *Berkeley: a new theory of vison and other writings* (London: Dent, 1972).

5. See D. Diderot, *Lettre sur les aveugles à l'usage de ceux qui voient* (1749; Paris: Garnier-Flammarion, 1972), p. 109. It is now known that Molineux's question is, for purely practical reasons, a bad one. Those who are congenitally blind and then undergo surgery (particularly delayed surgery) have a persistent visual deficit which includes an inability to distinguish shapes. See M. Jeannerod, "Déficit visuel persistant chez les aveugles-nés opérés," *Année Psychol.* 75 (1975): 169–196.

6. H. Dolezal, *Living in a world transformed: perceptual and performatory adaptation to visual distortion* (New York: Academic Press, 1982).

7. Dolezal's work contains many other interesting observations, which are, however, beyond the scope of this work. In particular, the author insisted upon the difficulty he experienced in reorganizing a coherent image of his own body. At the beginning of the experiment, when he saw the objects around him inverted, he also had the impression that his head was inverted, and found himself with his chin pointing upwards, his head between his knees. He also found that when he looked at his feet, he had the impression that they were above his head. Other parts of his body, seen inverted, seemed strange; he was often "surprised" to see his own hand in the course of a movement.

8. Held utilized a protocol very similar to Helmholtz's. See M. Jeannerod and H. Hécaen, *Adaptation*, chap. 6, "Adaptation and visuo-motor coordination."

9. See R. y Cajal, *Histologie des systèmes nerveux de l'homme et des vertébrés* (Paris: Maloine, 1909–1911; rpt. 1972).

10. D. O. Hebb, *The organization of behavior, a neuropsychological theory* (London: Chapman and Hall, 1949).

11. D. Hubel and T. Wiesel, laureates (with R. Sperry) of the Nobel Prize in 1981, had a determining influence in the renewal of nativism as an establishing force in the performance of neural networks. They situate themselves, however, in a context which reaffirms the predominance of hereditary factors in psychology.

12. This debate is nicely treated in M. Piattelli-Palmarini's *Language frmtand learning: the debate between Jean Piaget and Noam Chomsky* (Cambridge: Harvard University Press, 1980).

13. See J. Piaget on this subject, *Biologie et connaissance* (Paris: 1967). Piaget rejects the idea of a fixed, innate nucleus as proposed by Chomsky: "If reason is innate, either it is general and one must have it go back as far as the protozoa, or it is specific . . . and one must explain . . . through

which mutations and under the influence of which natural selections it developed" (in Piattelli-Palmarini, *Language*, p. 30).

14. *Biologie et connaissance*, pp. 14–15.

15. J. Prygogine and I. Stengers, in their work *La nouvelle alliance* (Paris: Gallimard, 1979), explain how certain complex entities may, when they find themselves in a state far removed from their point of equilibrium, acquire new properties. These "dissipative" structures can thus dissipate in an irreversible fashion the energy they contain. Thus, one can see a physical explanation for the idea of spontaneity.

16. E. du Bois-Reymond, *Uber die Grenzen des Naturkennens* (Leipzig, 1872). Cited by P. L. Assoun, *Introduction*, pp. 68–69.

Index